THE DUSTY UNIVERSE

THE ELLIS HORWOOD LIBRARY OF SPACE SCIENCE AND SPACE TECHNOLOGY

SERIES IN ASTRONOMY

Series Editor: JOHN MASON B.Sc., Ph.D.
Consultant Editor: PATRICK MOORE C.B.E., D.Sc.(Hon.)

This series aims to coordinate a team of international authors of the highest reputation, integrity and expertise in all aspects of astronomy. It makes valuable contributions to the existing literature, encompassing all areas of astronomical research. The titles will be illustrated with both black and white and colour photographs, and will include many line drawings and diagrams, with tabular data and extensive bibliographies. Aimed at a wide readership, the books will appeal to the professional astronomer, undergraduate students, the high-flying 'A' level student, and the non-scientist with a keen interest in astronomy.

THE MOON: Volume 1: Physics, Geology and Evolution
David Baker, Space Science and Engineering Consultant, Oundle, Northamptonshire
THE MOON: Volume 2: Exploration and Resources
David Baker, Space Science and Engineering Consultant, Oundle, Northamptonshire
THE AURORA: Sun–Earth Interactions
Neil Bone, British Astronomical Association, School of Biological Sciences, University of Sussex
PLANETARY VOLCANISM: A Study of Volcanic Activity in the Solar System
Peter Cattermole, formerly Lecturer in Geology, Department of Geology, Sheffield University, UK, now Freelance Writer and Consultant and Principal Investigator with NASA's Planetary Geology and Geophysics Programme
DIVIDING THE CIRCLE: The Development of Critical Angular Measurement in Astronomy 1500–1850
Allan Chapman, Centre for Medieval and Renaissance Studies, Oxford, UK
SATELLITE ASTRONOMY: The Principles and Practice of Astronomy from Space
John K. Davies, Royal Observatory, Edinburgh, UK
THE ORIGIN OF THE SOLAR SYSTEM: The Capture Theory
John R. Dormand, Department of Mathematics and Statistics, Teesside Polytechnic, Middlesbrough, UK, and Michael M. Woolfson, Department of Physics, University of York, UK
THE DUSTY UNIVERSE
Aneurin Evans, Department of Physics, University of Keele, UK
ASTEROIDS: Their Nature and Utilization
Charles T. Kowal, Space Telescope Institute, Baltimore, Maryland, USA
COMET HALLEY – Investigations, Results, Interpretations
Volume 1: Organization, Plasma, Gas
Volume 2: Dust, Nucleus, Evolution
Editor: J.W. Mason, B.Sc., Ph.D.
ELECTRONIC AND COMPUTER-AIDED ASTRONOMY: From Eyes to Electronic Sensors
Ian S. McLean, Joint Astronomy Centre, Hilo, Hawaii, USA
URANUS: The Planet, Rings and Satellites
Ellis D. Miner, Jet Propulsion Laboratory, Pasadena, California, USA
THE PLANET NEPTUNE
Patrick Moore, CBE, D.Sc.(Hon.)
ACTIVE GALACTIC NUCLEI
Ian Robson, Director of Observatories, Lancashire Polytechnic, Preston, UK
ASTRONOMICAL OBSERVATIONS FROM THE ANCIENT ORIENT
Richard F. Stephenson, Department of Physics, Durham University, Durham, UK
EXPLORATION OF TERRESTRIAL PLANETS FROM SPACECRAFT: Instrumentation, Investigation, Interpretation
Yuri A. Surkov, Chief of the Laboratory of Geochemistry of Planets, Vernandsky Institute of Geochemistry, Russian Academy of Sciences, Moscow, Russia
THE HIDDEN UNIVERSE
Roger J. Tayler, Astronomy Centre, University of Sussex, Brighton, UK
AT THE EDGE OF THE UNIVERSE
Alan Wright, Australian National Radio Astronomy Observatory, Parkes, New South Wales, Australia, and Hilary Wright

THE DUSTY UNIVERSE

ANEURIN EVANS
Department of Physics
University of Keele

JOHN WILEY & SONS
Chichester · New York · Brisbane · Toronto · Singapore

In association with
PRAXIS PUBLISHING LTD
Chichester

Reprinted May 1994

Wiley Offices:
John Wiley & Sons Ltd, Baffins Lane, Chichester,
West Sussex PO19 1UD, England

John Wiley & Sons, Inc., 605 Third Avenue,
New York, NY 10158-0012, USA

Jacaranda Wiley Ltd, GPO Box 859, Brisbane,
Queensland 4001, Australia

John Wiley & Sons (Canada) Ltd, 22 Worcester Road,
Rexdale, Ontario M9W 1L1, Canada

John Wiley & Sons (SEA) Pte Ltd, 37 Jalan Pemimpin #05-04,
Block B, Union Industrial Building, Singapore 2057

Orders & Returns:
The Distribution Centre,
Southern Cross Trading Estate,
1 Oldlands Way,
Bognor Regis PO22 9SA U.K.

Contents

Preface x

Introductory reading xii

1 Introduction 1

1.1 Why dust? 1
1.2 Units 4
1.2.1 Distance 4
1.2.2 Solar units 4
1.3 Quantifying the 'brightness' of an object 5
1.3.1 Flux and brightness 5
1.3.2 The magnitude scale 8
1.3.3 Colour indices 12
1.3.4 Absolute magnitudes 13
1.4 Stars 14
1.4.1 Stellar properties 14
1.4.2 Stellar evolution 16
1.4.3 Stellar populations 18
1.4.4 Variable stars 19
1.5 The Galaxy and the interstellar medium 19
1.5.1 The interstellar gas 20
1.5.2 The interstellar magnetic field 21
1.6 Gathering information about dust 22
1.6.1 Millimetre wavelengths 22
1.6.2 Infrared 23
1.6.3 Optical 24
1.6.4 Ultraviolet 26
1.6.5 X-ray 27

2 From astronomy to physics **29**
2.1 Introduction 29
2.2 The transfer of radiation 29
2.2.1 The equation of transfer 29
2.2.2 Solving the equation of transfer 32
2.3 Equivalent width 34

3 The interaction of a grain with radiation **39**
3.1 General ideas 39
3.2 Phonon modes in solids 39
3.2.1 Acoustic modes 39
3.2.2 Optical modes 41
3.3 The electrical properties of solids 44
3.4 Absorption, extinction and scattering 46
3.4.1 Radiation pressure 48
3.4.2 The variation of the Q-factors 49
3.4.3 Polarization 53
3.4.4 Elongated grains 55

4 Properties of dust particles **59**
4.1 Introduction 59
4.2 Grain charge 60
4.2.1 The photoelectric effect 60
4.2.2 Thermionic effect 61
4.2.3 Grain charge in a plasma 62
4.3 Temperature 65
4.3.1 Grain heating by radiation 65
4.3.2 Grain heating by gas impact 69
4.3.3 Grain heating by chemical reaction 74
4.3.4 Do grains have an equilibrium temperature? 76
4.4 Viscous drag 78
4.5 Poynting-Robertson effect 79

5 Grain formation and destruction **83**
5.1 Introduction 83
5.2 Change of phase 83
5.3 Grain nucleation 86
5.4 Grain condensation temperatures 89
5.5 Grain growth 89
5.5.1 Rate of grain growth 89
5.5.2 The effect of depletion of the condensing species . . . 91

5.5.3 Nature of grain 94
5.6 Grain destruction 95
5.6.1 Evaporation 95
5.6.2 Sputtering 96
5.6.3 Chemical sputtering 97
5.6.4 Grain-grain collisions 98
5.6.5 Other grain destruction mechanisms 99

6 Interstellar dust **103**
6.1 Introduction 103
6.2 The extinction law 103
6.2.1 The extinction 103
6.2.2 The extinction in magnitudes 104
6.2.3 Colour excess 105
6.2.4 The ratio of total-to-selective extinction 106
6.3 The wavelength-dependence of extinction 106
6.3.1 Extension beyond the optical 106
6.4 Composition of interstellar dust 111
6.4.1 Implications for grain size 111
6.4.2 Constraints from the gas-to-dust ratio 113
6.4.3 Depletion of elements 113
6.4.4 Dust-related absorption-features 115
6.4.5 The composition and size distribution of interstellar dust 116
6.5 Scattering by interstellar dust 117
6.5.1 Reflection nebulae 117
6.5.2 Scattering of background starlight by interstellar dust 120
6.5.3 Scattering when the illuminating source is variable . . 121
6.5.4 Compact reflection nebulae 127
6.5.5 X-ray scattering 129
6.6 Emission by interstellar dust 136
6.6.1 The temperature of interstellar grains 136
6.7 Polarization by interstellar dust 139
6.7.1 The alignment of interstellar grains 143
6.7.2 The relation between polarization and extinction . . . 146
6.8 The formation of interstellar dust 147
6.9 Mantle formation 148

7 Circumstellar dust **153**
7.1 Introduction 153
7.2 What is circumstellar dust? 153
7.3 Stars with infrared excess 155
7.3.1 Infrared two colour diagrams 155
7.3.2 Luminosity of the dust 158
7.3.3 Flux distribution of circumstellar dust 159
7.4 Spectral signatures of circumstellar dust 162
7.4.1 Oxygen-rich giants 164
7.4.2 Carbon stars 167
7.4.3 Pre-main sequence stars 167
7.5 Grain formation and growth 168
7.5.1 The nucleation problem 168
7.5.2 Grain growth in a stellar wind 170
7.5.3 Stars undergoing sporadic grain formation 173

8 The PAH hypothesis **179**
8.1 Introduction–How small can a solid be? 179
8.2 Anomalies in the infrared emission of interstellar dust 180
8.3 The nature of the small grains 182
8.3.1 Polycyclic aromatic hydrocarbons (PAHs) 182
8.4 Absorption and emission by PAH 182
8.4.1 Internal conversion 183
8.4.2 Temperature 184
8.5 Physical properties of PAH 188
8.5.1 Dimensions 188
8.6 The relation between PAH and larger grains 190
8.7 Problems with the PAH hypothesis 191
8.7.1 The infrared emission bands 192
8.7.2 Ultraviolet features 192
8.7.3 Laboratory and cosmic PAHs 194
8.8 Large carbon molecules 194

9 Extragalactic dust **197**
9.1 Introduction 197
9.2 Galaxies 197
9.2.1 The Hubble sequence 197
9.2.2 Other extragalactic objects 198
9.2.3 Clustering of galaxies 200
9.3 Some cosmological preamble 200
9.3.1 The expansion of the Universe 200

9.3.2 The redshift 201
9.3.3 The '3K' background 202
9.4 Dust in galaxies 203
9.4.1 Extinction in other galaxies 204
9.4.2 Infrared evidence for dust in other galaxies 207
9.5 Intergalactic dust 209
9.5.1 Intergalactic 'footballs' 209
9.5.2 Search for intergalactic dust 212

A Equilibrium charge on a spherical grain 215

B Grain size distribution from grain-grain shattering 219

C Partition function for a PAH molecule 223

D Astronomical co-ordinate systems 225
D.1 The celestial equator 225
D.2 Ecliptic co-ordinates (λ, β) 225
D.3 Galactic co-ordinates (l, b) 226

E Useful data 227
E.1 Constants 227
E.1.1 Physical constants 227
E.1.2 Astronomical constants 227
E.2 Useful Formulae 228
E.2.1 Blackbody formulae for temperature T 228
E.2.2 Maxwell speed distribution 228
E.2.3 Mathematical formulae 228

Index 231

Preface

It is common now in undergraduate Physics courses to include, at some point, optional courses in one or more areas of Astrophysics. Such courses provide excellent examples of the applications of Physical principles, taught in the 'core' Physics course, in areas that appeal to many undergraduates. Furthermore they often provide the opportunity of teaching areas of Physics that may be omitted from the core course. One drawback is that there are very few texts that are suitable for such courses. While there are many excellent texts that would be suitable for undergraduate astronomers these usually assume a background knowledge of astronomical essentials that most Physics undergraduates do not possess. This book, the contents of which are based on courses given by the author in the Physics Department at Keele, is therefore meant to serve the needs of undergraduate Physics students who may be following a course on various aspects of cosmic dusts, either as an extension of a Solid State Physics course or as an optional astrophysics course. The book is intended to provide the basics; since the subject matter is evolving at a rapid pace any course that uses this material should supplement the text with the results of more recent research.

When I was making the initial plans for the contents, I had intended to include a chapter on dust in the Solar System. However as the writing progressed and the amount of material grew it became apparent that, to do justice to Solar System dust, the book would have been nearly twice as long as it now is. I therefore made the decision to omit it altogether, although reference is made at several points to solar system solids.

A book like this can certainly make no claim to originality; the contents can be found in many of the texts and review papers referred to at the ends of the individual chapters. M. F. Bode, J. Mason, W. J. Pigram, T. Naylor and D. A. Williams read and made constructive comments on some or all of early versions of the manuscript. They have all done their best to draw my attention to ways of improving the presentation and clarity; I hope that they will feel that their efforts have not been in vain. The editorial staff of Ellis Horwood were especially vigilant; their careful scrutiny of the manuscript revealed several typographical errors that I had missed. It goes without saying that the errors and omissions that now remain are my own responsibility.

I extend my gratitude to the many authors, editors and institutions who have granted permission to reproduce figures from their published work, and to Mr M. Daniels for his help in preparing the figures for publication. The manuscript was prepared using the LaTeX document preparation software and Mr G. Pratt was helpful in guiding me down some of the murkier backwaters of TEX.

I would finally like to express my thanks to my wife Michelle, who suffered the writing of this book more than I did, and to the officers of Ellis Horwood, who readily took this venture on, and who were very patient as one deadline after another passed with no manuscript forthcoming.

March 24, 1993

A. Evans

Introductory reading

The list of general texts below is not meant to be exhaustive, and reflects to some extent the prejudices and preferences of the author. Reference to material relevant to the contents of individual chapters is given at the end of each chapter; these references include not only standard texts but also, where appropriate, review papers in the research literature. Throughout, an indication of the level expected is given, as follows:

[A]–written for the well-informed lay-person, with no sophistication in astronomy, physics or mathematics required.

[B] - knowledge of astronomy/physics/mathematics necessary, at the level of the sixth form/first year university.

[C] - undergraduate level.

[D] - graduate level.

The various 'Cambridge' books on astronomy, such as those listed below, are lavishly illustrated and written by authorities in their respective fields:

[A] *The Cambridge Atlas of Astronomy*, Cambridge University Press (1985).

[A] *The Cambridge Encyclopaedia of Astronomy*, Cambridge University Press (1985).

An excellent introduction to general astronomy, superbly illustrated, is given in

[A] The Dynamic Universe, T. P. Snow, West Publishing Company (1991).

A discussion at a somewhat more advanced level is given in the following:

[B] *Astronomy–A Physical Perspective*, M. L. Kutner, Harper & Row (1987).

The following texts are also at a more advanced level; the first contains excellent discussions not only of astrophysical concepts (as the title implies) but also of some important physical ideas, while the second provides a useful discussion of radiation processes. The third is a useful compendium of useful formulae etc.:

[C] *Astrophysical Concepts*, M. Harwit, Springer Verlag (1988).

[C/D] *Radiative Processes in Astrophysics*, G. B. Rybicki & A. P. Lightman, J. Wiley (1980).

[C/D] *Astrophysical Formulae*, K. R. Lang, Springer Verlag (1974).

1

Introduction

1.1 Why dust?

When one looks at the night sky the presence of small dust particles in large quantities is not exactly the first thing that grasps the attention. Why then is so much of modern astronomy and astrophysics–and much effort in terrestrial laboratories–devoted to the study of this apparently insignificant component of the Universe?

When William Herschel conducted his visual survey of the stars in the 18 th century, he noted the presence of 'holes' in the distribution of stars on the sky (see Fig. 1.1). He suggested that these holes might indeed be holes in the sky and it was not until the early 20 th century that astronomers began to suspect that these holes are in fact regions of heavy obscuration. Conclusive evidence that there is obscuring material in interstellar space was obtained in 1930 by (amongst others) Trumpler, who essentially measured the distances of clusters of stars using two independent methods. The first involved measuring the apparent luminosities S of stars in a cluster, the second involved measuring the apparent diameter ϕ of the cluster as a whole. If one knows the intrinsic properties of the cluster, in this case the ***absolute*** luminosities L of its stars and its linear dimensions d, one can determine its distance D from the relations [see Eq. (1.15) below]

$$S = \frac{L}{4\pi D^2} \qquad\qquad \phi = \frac{d}{D}.$$

Trumpler found that cluster distances as determined using the luminosities of the stars were systematically greater that those determined using their dimensions and, moreover, the discrepancy between the two increased with increasing distance. He concluded that the space between the stars is not empty, and that there is a component of the interstellar medium that absorbs

Figure 1.1: Milky Way in Sagittarius, showing clouds of obscuring dust. Photograph from the Hale Observatories.

starlight. Note that the presence of obscuring material makes objects appear fainter (and therefore further away) than they really are but has no effect on their apparent size.

One can imagine the collective cry of anguish from astronomers who, at that time, were just beginning to understand the physical processes occurring in stellar interiors (which determine the structure of stars) and in stellar atmospheres (which determine their observational properties). The presence of this obscuring material was clearly a great nuisance whose presence prevented a straightforward study of the stars themselves.

A major effort was then initiated to investigate the properties of this material, not as something worthy of study in its own right but as a means of removing its effects from the observations of stars. Early work, in the 1940's, measured the way in which the obscuration varied with wavelength and it was at that time that the nature of the obscuring material became apparent. In the visible the obscuration varies with wavelength λ approximately as λ^{-1} and this immediately identifies the obscuring material as small, solid particles having dimensions $\sim$ an optical wavelength (although at this stage it says nothing about either the geometry or the chemical composition of the particles).

It was the emergence of infrared astronomy as an important observational tool that clinched the importance of these annoying little dust particles as significant constituents of the Universe and indeed, as vital ingredients in several poorly-understood astrophysical processes. They are implicated as essential participants in the early stages of star formation; they are necessary for the formation of molecules in interstellar space; they can drive the mass loss that occurs when a star is approaching the end of its life; and they conspire to form planets around stars like the Sun, on which curiosity-driven species eventually evolve and attempt to understand the Universe around them. It is ironic that the material which astronomers felt was such a hindrance to the study of stars turns out to be a fundamental catalyst in the formation and final evolution of the stars themselves!

The study of dust particles in astrophysical environments is now a major area in astrophysics and furthermore, presents formidable challenges to the ingenuity of physicists (and chemists) in understanding their fundamental properties. Virtually every area of physics is necessary to understand the physical properties of cosmic dust grains. To understand how they interact with radiation, by absorbing and scattering, one must appeal to solid state physics and to electromagnetic theory. An understanding of the heating of interstellar grains immersed in a hot gas, or by chemical reactions on grain surfaces, comes from surface physics. We again look to solid state physics to understand why dust grains might be electrically charged, while the heating of grains, especially small grains, is a problem in statistical physics and thermodynamics. The physics and chemistry of small atomic clusters, in which there is considerable interest in terrestrial laboratories, is also of relevance to the astrophysics community. Such clusters, having properties that are unattainable on Earth, exist in astronomical environments either as nucleation sites during the early stages of grain growth in stellar atmospheres, or even as components of the interstellar grain population. Indeed the small dust particles that populate the space between the stars present the terrestrial physicist with the opportunity of studying a state of matter that can not be attained in the laboratory.

The investigation of dust particles in astronomical environments is one of the most challenging and exciting areas of current astronomy and we aim in this book to provide an introduction to the basic physics of dust grains. However, before we proceed to look at the physics of dust particles in interstellar and circumstellar space we devote the remainder of this chapter to provide some pertinent astronomical background.

1.2 Units

The nature of astronomy has meant that a system of units unique to the subject has evolved over the years and we discuss these in this section. However, the units used for measuring the 'brightness' of an object merit a more extensive discussion and we defer this to the following section. We should note that most of the astronomical literature makes use of the c.g.s. system of units, although some research journals are slowly moving over to the S.I. system. While this generally poses no great difficulty when we discuss quantities involving length, time and their derivatives etc., problems are more likely to arise when electromagnetic quantities are encountered. In this book we shall generally use the S.I. system.

1.2.1 Distance

It is no accident that the adjective 'astronomical' has evolved, by common usage, to mean 'extremely large' and the magnitudes of astronomical distances are typical of the large numbers encountered in the subject.

The Earth moves around the Sun in an elliptical orbit, the semi-major axis of which has been accurately measured to be 149.6×10^6 km. This distance is commonly referred to as the *Astronomical Unit*, or AU and this is commonly used as the unit of distance for objects within the solar system.

The light year–the distance (9.46×10^{15} m) a photon travels in a year *in vacuo*–is commonly encountered in popular accounts of astronomy; it is a valuable concept in that it provides a 'handle' with which one can grasp the vastness of astronomical distance scales. However, it is rarely used in the research literature and the unit of distance that is normally employed is the *parsec* (abbreviated to pc), the distance at which the Astronomical Unit would subtend an angle of one second of arc. Thus

$$1 \text{ pc} = \frac{149.6 \times 10^6 \times 10^3}{\pi/(180 \times 3600)} = 3.09 \times 10^{16} \text{ m}.$$

A parsec is about $3\frac{1}{4}$ light years. The usual decimal prefixes, such as kpc for 10^3 pc and Mpc for 10^6 pc, are also commonly used. The scale of the observable Universe is about 4 Gpc.

1.2.2 Solar units

The Sun is a conveniently situated star whose parameters are easily measured. These parameters provide a convenient set of units for describing stellar quantities in general and set the scene for a description of stars in general in Section 1.4. The Sun's vital statistics are:

Luminosity: $L_\odot = 3.83 \times 10^{26}$ W;
Mass: $M_\odot = 1.99 \times 10^{30}$ kg;
Radius: $R_\odot = 7.00 \times 10^{8}$ m.

1.3 Quantifying the 'brightness' of an object

Astronomy, like the other sciences, has not developed in a logical fashion. Until the 20 th century, the optical was the only wavelength range capable of being exploited, and it wasn't until the 1930's that other windows on the Universe, beyond the optical, were opened. Infrared astronomy is a discipline in the development of which astronomers themselves took the initiative but radio astronomy got off the ground thanks largely to the efforts of radio engineers. The result is that the units used in various branches of astronomy to measure the 'brightness' of an object bear the imprints of the early pioneers.

1.3.1 Flux and brightness

Before discussing the systems of units in use, we should first clear up a question of definition: we should distinguish between the *brightness* of a star and its *flux*. When we look at the star with a detector (e.g. the eye), what we detect is the flow of radiant energy falling on the detector in unit time; thus we would measure the flux S in (for example) $\mathrm{W\,m^{-2}}$. However, detectors are usually sensitive in only restricted regions of the electromagnetic spectrum and we have to specify how the radiative energy is distributed in frequency ν (or, equivalently, in wavelength λ). This is done by measuring the flux (in $\mathrm{W\ m^{-2}}$) in as narrow a bandwidth as is convenient and feasible; for example, this can be done (as in the optical) by using special filters, or (as at radio wavelengths) electronically. If the frequency scale is in Hz, then the quantity we measure is the flux in unit frequency bandwidth, or the *flux density*. This is denoted by S_ν, the suffix ν emphasizing that it is the flux in unit frequency interval; the units of S_ν are $\mathrm{W\,m^{-2}\,Hz^{-1}}$. On the other hand if we are measuring the flux in wavelength bands, the flux density S_λ is in $\mathrm{W\,m^{-2}\,\mu m^{-1}}$ if the wavelength scale is in μm. It gives some idea of the faintness of astronomical sources when we note that a convenient unit for measuring the flux densities of the *brighter* sources is the *Jansky*, $10^{-26}\ \mathrm{W\,m^{-2}\,Hz^{-1}}$.

Note that, since we are measuring the flux density S_ν in a narrow frequency range $\delta\nu$, the total flux measured over the bandwidth $\delta\nu$ is $S_\nu\delta\nu$. This enables us to convert from 'frequency' units (in $\mathrm{W\ m^{-2}\ Hz^{-1}}$) to 'wavelength' units (in $\mathrm{W\ m^{-2}\ \mu m^{-1}}$). If the frequency interval $\delta\nu$ corresponds to the wavelength interval $\delta\lambda$ then we must have that

$$S_\nu \delta\nu = S_\lambda \delta\lambda \tag{1.1}$$

if $\delta\nu$ and $\delta\lambda$ are both narrow enough that the flux density does not vary appreciably within them. Now we know that

$$\lambda\nu = c, \tag{1.2}$$

where c is the velocity of light. Differentiating Eq. (1.2) we get:

$$\delta\nu = -\delta\lambda c/\lambda^2. \tag{1.3}$$

A little thought should convince the reader that we are not interested in the minus sign in Eq. (1.3), which tells us only that, as the frequency scale increases, the wavelength scale decreases; we are interested only in the relationship between the *magnitudes* of the bandwidths $\delta\nu$ and $\delta\lambda$. We can do this by taking Eq. (1.3) and ignoring the minus sign (i.e. we take the modulus). Combining Eqs. (1.1) and (1.3) we get

$$S_\nu = S_\lambda \lambda^2/c \tag{1.4}$$

or, equivalently

$$S_\lambda = S_\nu \nu^2/c, \tag{1.5}$$

where we measure the flux densities in 'frequency units' (e.g. W m^{-2} Hz^{-1}) and in 'wavelength units' (e.g. W m^{-2} μm^{-1}) respectively.

So much for flux and for flux density; how does brightness come into all this? To see this consider the photograph of the Orion nebula in Fig. 1.2. The large-scale structure of the object is resolved and there are quite clearly variations in brightness with position across the extent of the object. Now the total flux from the Orion nebula is obtained by adding together the contribution from each part of the nebula. Suppose we take a small element of an object (like the Orion nebula); such an element will subtend a specific solid angle at the observer (see Fig. 1.3). If the area of the element is dA, and its distance from the observer is D, then the solid angle $d\Omega$ subtended is given by

$$d\Omega = dA/D^2. \tag{1.6}$$

The usual units of Ω are *steradians*, although astronomers sometimes use square degrees or square arcseconds (there are $32400/\pi^2$ square degrees in a steradian). Now brightness Σ_ν is measured in units of flux density per unit solid angle (e.g. W m^{-2} Hz^{-1} sr^{-1}). We get the flux density by integrating

Figure 1.2: The Orion nebula. Photograph from the Hale Observatories.

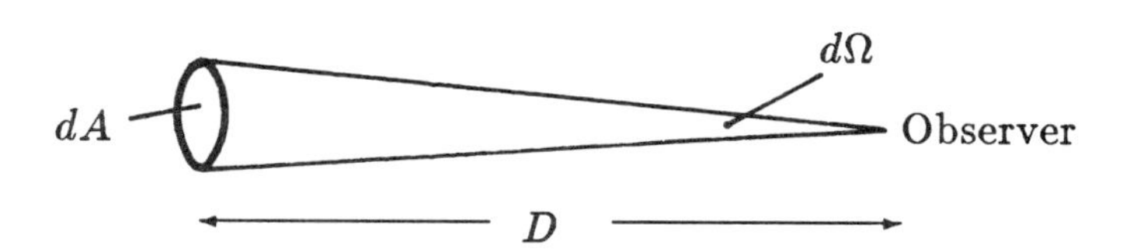

Figure 1.3: Definition of solid angle.

over solid angle; in other words, we add up the contribution to the flux from all the elements like dA:

$$S_\nu = \int_\Omega \Sigma_\nu d\Omega, \tag{1.7}$$

where the integration is carried out over the solid angle Ω subtended by the source. Note that the brightness Σ_ν is independent of distance and is essentially determined by whatever physical process is responsible for the emission of radiation. If the Orion nebula were somehow transported to the most distant galaxy, its observed brightness would be unchanged but the flux would [according to Eq. (1.7)] be greatly reduced by virtue of the reduced solid angle; indeed, by comparing Eqs. (1.6) and (1.7), notice how the familiar inverse square dependence of flux on distance drops out of Eq. (1.7) by virtue of the dependence of S_ν on $d\Omega$. Similarly, any solar-type star has the same brightness as the Sun (neglecting such complications as limb darkening etc.); the Sun has greater flux compared with other similar stars because of its proximity and therefore larger solid angle. If we approximate a star by blackbody emission at temperature T_* from a uniform circular disc of radius R_* at distance D, then Σ_ν is the Planck function $B_\nu(T_*)$ and $d\Omega = \pi R_*^2/D^2$. In this case

$$S_\nu = \int_\Omega \Sigma_\nu d\Omega = B_\nu(T_*)\frac{\pi R_*^2}{D^2}. \tag{1.8}$$

1.3.2 The magnitude scale

In the optical, the scheme for measuring stellar flux goes back over twenty centuries. The Greek astronomer Hipparchus (190–120 B.C.) compiled the first catalogue of stars which was, for obvious reasons, based on naked eye observation. Hipparchus classified the stars according to their flux. The brightest stars visible to the naked eye (such as Sirius, Antares etc.) Hipparchus referred to as 'stars of the first magnitude'; those that were slightly less 'bright' he referred to as 'stars of the second magnitude', and so on until the 'stars of the sixth magnitude' were the faintest visible with the naked eye.

The classification according to flux is clearly extremely crude; indeed even a casual glance at the night sky shows that not only do Hipparchus' 'first magnitude stars' not have the same flux, but also that they have different colours. However, Hipparchus' classification has, in essence, remained and in the 19 th century it was found that 6 th magnitude stars were roughly 100 times fainter than 1 st magnitude stars. The magnitude scale was therefore redefined in such a way that a *difference* of five magnitudes is precisely equivalent to a *flux ratio* of one hundred. Thus a 6 th magnitude star is exactly 100 times fainter than a star of 1 st magnitude, while a star of 12 th

Table 1.1: Effective wavelengths of optical and infrared filters; units of $[S_\lambda]_0$ are W m^{-2} μm^{-1}

Optical				Infrared			
Filter	$\lambda_{\rm eff}$ (μm)	$\Delta\lambda$ (μm)	$\log[S_\lambda]_0$	Filter	$\lambda_{\rm eff}$ (μm)	$\Delta\lambda$ (μm)	$\log[S_\lambda]_0$
				J	1.25	0.35	-8.50
U	0.36	0.07	-7.37	H	1.65	0.35	-8.93
B	0.44	0.10	-7.18	K	2.2	0.4	-9.38
V	0.55	0.09	-7.44	L	3.5	0.8	-10.21
R	0.70	0.22	-7.76	M	4.8	0.6	-10.68
I	0.90	0.24	-8.08	N	10	0.8	-11.91
				Q	20	0.10	-13.14

magnitude is 100 times fainter than a 7th magnitude star, and so on. In addition, each magnitude interval is subdivided, and the scale is extended to include negative numbers so that the brightest stars and planets can be incorporated in the scheme.

Mathematically we can relate the flux S to the magnitude m by the formula

$$m = -2.5 \log_{10} S + \text{constant}, \tag{1.9}$$

where the value of the constant has to be determined (i.e. the magnitude scale has to be calibrated) by observing stars the fluxes of which have been determined by other methods. Note that, on this scale, the *brighter* stars have *smaller* magnitude numbers, a fact that often confuses newcomers to the subject.

Clearly the magnitude scale, as discussed so far, is not really adequate. It would be nice to be able to specify the magnitude of a star in different wavelength ranges, e.g. in the red or blue. Over the years an internationally agreed set of standard broad-band filters has been devised, so that astronomers at different observatories measure the same thing when they observe the same star. In the optical these filters are centred on wavelengths listed in Table 1.1. The response of the *UBV* filters as a function of wavelength is shown in Fig. 1.4 and these form part of the *Johnson UBVRI* system, after H. L. Johnson, a pioneer in the field of stellar photometry. In Table 1.1 and Fig. 1.4, U stands for ultraviolet, B for blue, V for visual (yellow), R for red and I for 'infrared'. Infrared here means that part of the electromagnetic spectrum just beyond the red and actually before the infrared proper (which most astronomers generally take to begin at a wave-

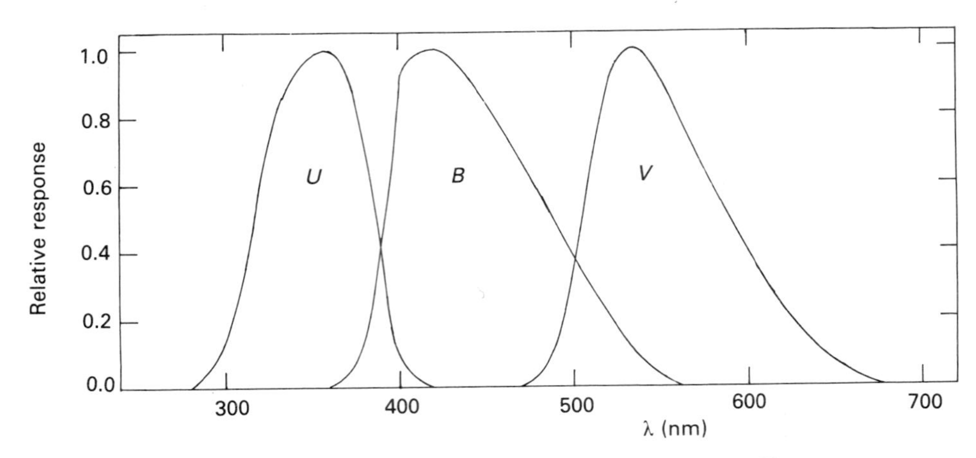

Figure 1.4: Wavelength response of *UBV* filters.

length of about 1μm). Also given in Table 1.1 is the approximate width $\Delta\lambda$ (in μm) of each filter.

A similar set of filters has been developed for use in the near and mid infrared (see Table 1.1), although it should be noted that there are often small but important differences between actual effective wavelengths and those listed in Table 1.1. Their response as a function of wavelength is shown in Fig. 1.5. In general, observations using such broad-band filters tend to 'smear out' much spectral information: the *spectral resolution* $\lambda/\Delta\lambda$ is poor. On the other hand data acquisition using broad-band filters is relatively rapid and the observer has to decide whether the observational programme requires coarse data for a large number of objects or detailed data for fewer objects.

Since the results of optical and infrared photometry are given in magnitudes, we need a means of converting from magnitudes to flux density. The essence of this is of course given by Eq. (1.9) but we need to refine this to take into account the wavelength response of the filter being used.

Suppose that the response of filter X (where X might be U, B, V, J, H, K etc.) as a function of wavelength is $\Phi_X(\lambda)$ at some wavelength λ, and that the flux density of the source has the value S_λ at the same wavelength. In general, $\Phi_X(\lambda)$ will include not only the response of the filter itself but also the wavelength-dependence of the atmospheric transmission and of the reflectivity of the telescope optics etc. The flux actually measured in a narrow wavelength interval $\delta\lambda$ is given by $\Phi_X(\lambda)S_\lambda\delta\lambda$. Thus the *total* flux measured through the filter, summed over all wavelengths, is given by integrating over

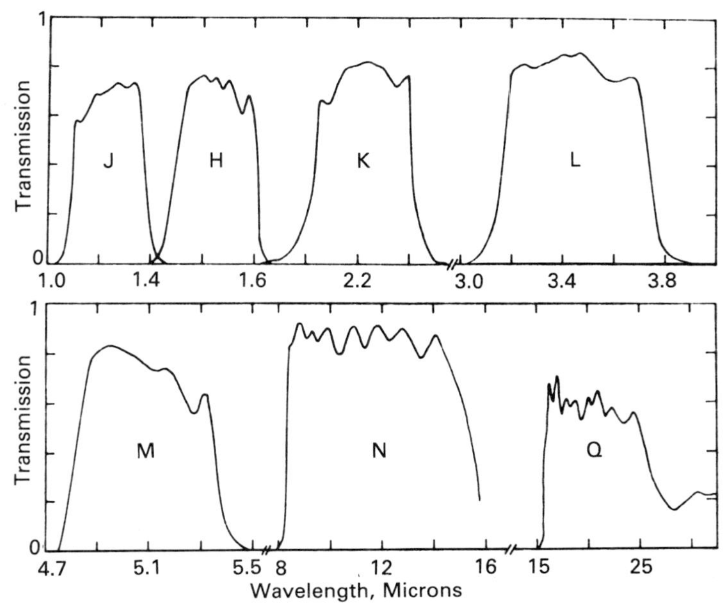

Figure 1.5: Wavelength response of infrared filters. After I. S. Glass, *Monthly Notices of the Royal Astronomical Society*, Vol. **164**, 155 (1973).

the entire bandwidth of the filter:

$$S_{\mathrm{X}} = \int_0^\infty \Phi_{\mathrm{X}}(\lambda) S_\lambda d\lambda, \tag{1.10}$$

where the suffix X on S_{X} reminds us that we are talking about the total flux through filter X. By analogy with Eq. (1.9) we can therefore write

$$m_{\mathrm{X}} = -2.5 \log_{10} S_{\mathrm{X}} + \text{constant}. \tag{1.11}$$

Note that the value of the constant also depends on whether the flux density is measured in wavelength or frequency units. Thus we end up with the magnitudes m_{U}, m_{B}, m_{V} etc. or, as they are more commonly known, U, B, V etc. The flux density corresponding to zero magnitude for each filter is given in Table 1.1, so that the constant in Eq. (1.9) is determined.

One word of caution. Despite the fact that there is an agreed set of filters there are inevitable differences which arise largely as a result of local conditions; in particular sources having emission line spectra can be extremely troublesome. Strong emission lines around the wavelengths where the filter response falls steeply (see Figs. 1.4, 1.5) can have a significant effect on the measured magnitudes and photometric systems at different observatories may or may not include such lines in the measured magnitude. Accordingly

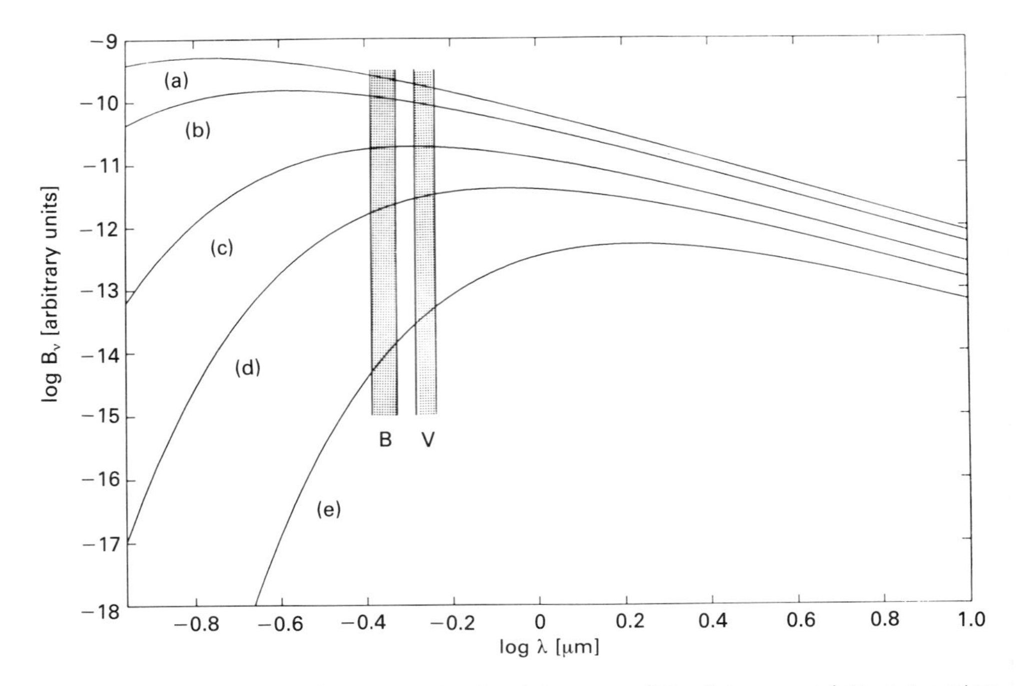

Figure 1.6: Planck function B_ν for (a) 3×10^4 K, (b) 2×10^4 K, (c) 10^4 K, (d) 6×10^3 K, (e) 3×10^3 K. Vertical strips indicate approximate locations of B and V filters.

it is generally unwise to directly compare magnitudes obtained at different observing sites.

1.3.3 Colour indices

It is frequently useful to define a quantity known as the *colour index*, which is the difference between two magnitudes at different wavelengths, for example $(B - V)$, $(U - B)$, $(K - L)$. Suppose we could observe a perfect blackbody (indeed for our present purposes we can regard stars as being good approximations to blackbody radiators). From blackbody physics, the distribution of the brightness of a blackbody amongst the various wavelengths is determined only by the temperature T of the emitting body, according to Planck's radiation law:

$$B_\nu(T) = \frac{2h\nu}{c^2} \frac{1}{\exp[h\nu/kT] - 1}. \tag{1.12}$$

Fig. 1.6 shows the dependence of $B_\nu(T)$ on wavelength for temperatures in the range 3000–30000 K, fairly typical of the range of stellar temperatures.

Two broad bands, taken to be representative of the B and V photometric filters, are also included in Fig. 1.6. It is evident that the relative amount of radiation in the B and V bands varies considerably for different values of T. Now from Eq. (1.11)

$$(B - V) = -2.5 \log \frac{S_{\rm B}}{S_{\rm V}} + \text{constant}, \tag{1.13}$$

so that $(B - V)$ is a direct measure of the relative amounts of radiation in the B and V bands; the colour index $(B - V)$ can therefore give a direct measure of the temperature of the radiating source. In fact

$$(B - V) = \frac{7300}{T} - 0.60. \tag{1.14}$$

A temperature derived in this way is referred to as the *colour temperature.* There are obviously equivalent relationships for $(K - L)$ etc. However, there is a complication as far as $(U - B)$ is concerned. For real stars, there is a 'step' in the flux distribution at a wavelength of 3648Å; at shorter wavelengths the stellar flux distribution may depart considerably from that of a blackbody because of absorption by atomic hydrogen.

As we shall see in Chapter 6, however, the observed $(B - V)$ is unlikely to give direct information about the temperature of a star. This is because interstellar (and circumstellar) dust increases the value of $(B - V)$ (i.e. *reddens* the star) so that, for a star of known temperature, the measured index does not correspond to that predicted by Eq. (1.14). However, if the intrinsic stellar $(B - V)$ is known from other observations, the difference between the observed and intrinsic $(B - V)$ tells us something about the intervening material.

1.3.4 Absolute magnitudes

The magnitude system discussed so far refers to measured values, i.e. apparent magnitudes corresponding to observed flux densities. The observed flux clearly depends both on distance and on absolute luminosity. We can also define absolute magnitudes to describe absolute luminosities. We have already seen how flux density falls with the inverse square of the distance. For an object of luminosity L at distance D conservation of energy gives the flux as

$$S = \frac{L}{4\pi D^2}, \tag{1.15}$$

provided the source radiates isotropically. Taking logs to base 10 of both sides of Eq. (1.15) and multiplying by -2.5 we get:

$$m = -2.5 \log_{10} L + 5 \log_{10} D + \text{constant}. \tag{1.16}$$

The absolute magnitude M is defined as the apparent magnitude an object would have if it were placed at a standard distance. In other words, in order to compare the absolute luminosities of objects we place them all at the same distance, which is chosen (quite arbitrarily) to be 10 pc. With this definition Eq. (1.16) becomes

$$m = M + 5\log_{10} D - 5, \tag{1.17}$$

where $M = -2.5\log_{10} L$+constant. However, Eq. (1.17) does not take into account the fact that starlight is diluted not only by the inverse square law, but also by material in interstellar space and Eq. (1.17) has to be modified accordingly.

Obviously there is an absolute magnitude that corresponds to each of the apparent magnitudes described in the previous section. From Eq. (1.17) the quantity $m - M$ depends only on the distance D of the object and is sometimes referred to as the *distance modulus.*

1.4 Stars

As we shall see in Chapters 6 and 7, there is good evidence for the hypothesis that most (if not all) dust in the Universe originates in stellar atmospheres. The conditions under which this can occur will be discussed in greater detail later but in general, we note that stars which tend to have large amounts of dust associated with them tend to be at the extreme ends of the stellar evolutionary scale. Since we shall be discussing the origin of cosmic dust and the properties of dusty stars in some detail in later chapters, we give here a brief description of the way in which stars are classified.

1.4.1 Stellar properties

A star may be characterized by two fundamental parameters, namely its bolometric luminosity $L_{\rm bol}$ and its temperature. The former is simply the total power emitted by the star, summed over all wavelengths. In general, the power emitted might include not only radiation, but also power in 'mechanical' form, for example in the form of a 'wind'; here however we shall neglect mechanical contributions to a star's power output. In general of course, observers can not determine $L_{\rm bol}$ directly, for the simple practical reason that they do not have access to the complete range of wavelengths. Instead they determine (for example) the emitted power $L_{\rm vis}$ at visual wavelengths, which is comparatively easy, and apply a *bolometric correction.* In terms of magnitudes,

$$M_{\rm bol} = M_{\rm vis} - B.C., \tag{1.18}$$

where the bolometric correction $B.C.$ is always positive using the definition (1.18).

The definition of stellar temperature is somewhat more troublesome. One measure of stellar temperature is given by the Stefan-Boltzmann law of blackbody physics, according to which the power emitted (in the form of radiation) by unit area of a blackbody at temperature T is given by σT^4. For a spherical star, of radius R_*, the total surface area is $4\pi R_*^2$ and so the total emission is

$$L_{\text{bol}} = 4\pi R_*^2 \sigma T_{\text{eff}}^4. \tag{1.19}$$

The temperature T_{eff} defined in this way is the *effective temperature* of the star. Another measure of temperature is provided by the wavelength at which the flux density is a maximum. The wavelength of maximum intensity λ_{max} is related to the temperature T_{Wien} via the Wien displacement law, again from blackbody physics:

$$\lambda_{\text{max}} T_{\text{Wien}} = 2890\,\mu\text{m K}, \tag{1.20}$$

where the wavelength λ_{max} is given in μm. Note that Eq. (1.20) refers to the intensity peak in B_λ; B_ν peaks at a different wavelength and for this case, the numerical constant in Eq. (1.20) should be replaced by 5100. Yet another measure of temperature is given by taking the ratio of fluxes at two wavelengths (for example blue and yellow–0.44 μm and 0.55 μm). Recall [Eq. (1.14)] that this is related to the colour index $(B - V)$. We then find the temperature of the blackbody that has the same flux ratio at these wavelengths. This measure of temperature is called the *colour* temperature T_{col}. If stars were perfect blackbody radiators, all three measures of temperatures would be the same. However, since stars are not blackbody radiators the three temperature measures differ. For example, in the case of the Sun, $T_{\text{eff}} = 5770$ K, $T_{\text{Wien}} = 6280$ K and the colour temperature $T_{\text{col}} = 5840$ K.

The stellar spectrum also gives an indication of its temperature because the temperature of a star near its 'surface' determines which elements are ionized, which particular atomic and ionic energy levels will be excited etc. It is customary to classify stellar spectra by a scheme of letters O,B,A,F,G,K,M. If the reader gets the impression that this is very nearly a random selection of letters, deliberately chosen to confuse, then this would be entirely understandable. In fact stellar spectra were first classified according to the strength of the hydrogen (Balmer) absorption lines in the spectra and the original classification was 'A' for stars having the strongest lines, 'B' for the next strongest and so on. As astronomers and physicists began to better understand the physical processes responsible for the strength of spectral lines it was realized that the sequence of stellar spectra is in fact a temperature sequence. When the classification was reorganized in order of decreasing temperature (but while maintaining the old letter classification) the letters became jumbled and some were dropped altogether; the result is the apparently random sequence of letters currently in use.

The letter classification for stellar spectra is reasonably adequate but obviously stars do not all fit neatly into seven temperature categories. The classification scheme is refined by dividing each letter up into ten subdivisions (O0–O9,B0–B9,A0–A9 etc,) and this is certainly adequate for most purposes. Thus for example the Sun has a G2 classification, whereas one of the brightest stars in Orion (Rigel) is classified B8. An extension of this classification scheme includes the spectral types R, N and S; these are cool, chemically peculiar, stars usually in an advanced state of evolution. The R and N type stars are carbon-rich and their spectra include features due to molecules containing carbon, such as C_2, CN; for this reason these stars are more commonly referred to as C-type stars. The C class is subdivided into C1, C2, C3..., depending on the strength of carbon features. The spectra of S-type stars, unlike those of M-type stars which contain TiO bands, contain instead ZrO bands.

1.4.2 Stellar evolution

A plot of $L_{\rm bol}$ against $T_{\rm eff}$ provides some useful pointers to stellar evolution. However, for convenience, it is more usual to plot the absolute visual magnitude $M_{\rm vis}$ against the colour index $(B - V)$; such a plot is essentially one of $L_{\rm vis}$ against colour temperature and is called a *Hertzsprung-Russell* (or H–R) diagram[1]. We could plot such a diagram for the stars in the solar neighbourhood, because the distances (and hence luminosities) of these stars are well determined; a plot for these stars is shown in Fig. 1.7. The most striking feature of Fig. 1.7 is the fact that most of the stars lie in a narrow band going from top left to bottom right; this is the *main sequence* and is effectively an expression of the Stefan-Boltzmann law of radiation [cf. Eq. (1.19)]. The main sequence on the H–R diagram does not quite reflect the Stefan-Boltzmann law, however, because of (for example) bolometric correction effects.

Also evident is the fact that there are some stars that lie to the upper right of the diagram and from their position on the diagram these stars are highly luminous and cool. Eq. (1.19) then shows that this can happen only if the stars are large: these are *red giant* stars. Similarly those stars that lie at the lower left are of low luminosity and high temperature; such stars

[1]The original H–R diagram was a plot of absolute visual magnitude against spectral type.

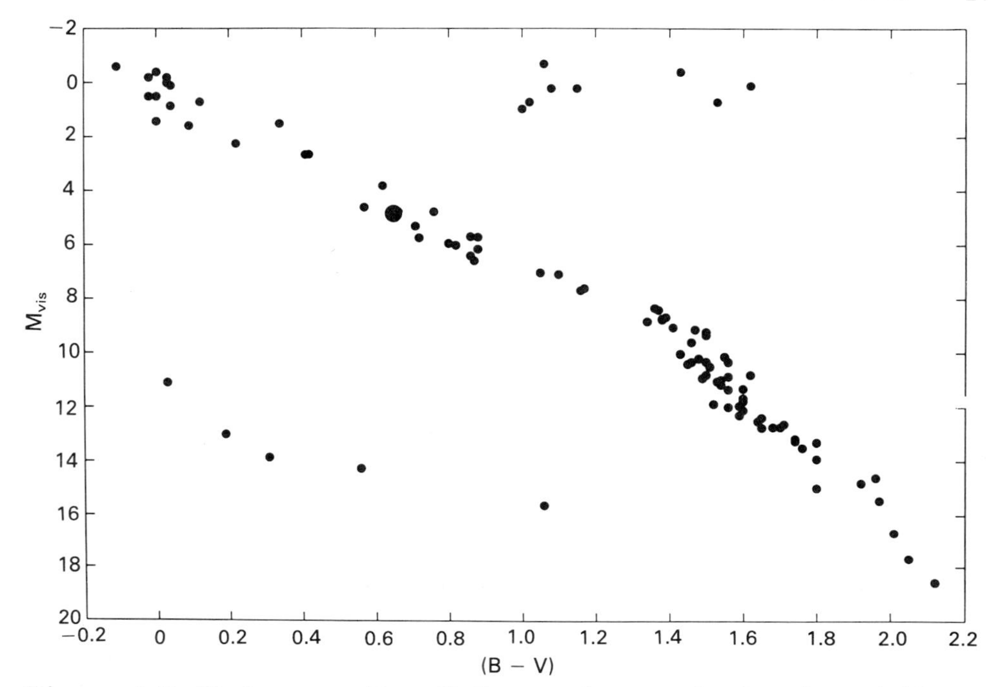

Figure 1.7: Hertzsprung–Russell diagram for stars in the solar neighbourhood. Large circle represents the Sun.

must, according to Eq. (1.19), be small and these objects are referred to as *white dwarf* stars.

As a star evolves, from initial formation to final demise, it will occupy different regions of the H–R diagram although generally speaking, a star will spend most of its life (more than 90 per cent) close to some point (determined by its mass) on the main sequence. Prior to a star settling on the main sequence (the *pre-main sequence* stage), and after it has left (the *evolved* state), the star will move around the H–R diagram in a way that is (again) determined by its mass.

At the time when the H–R diagram was first discussed in the 1920's it was believed that the hotter stars were at an early evolutionary stage, whereas cooler stars were well advanced in their evolution. While we realize now that such an idea is inaccurate it is unfortunate that the nomenclature 'early' and 'late' for O,B,A and K,M type stars respectively, has stuck. This is yet another example of the way in which a misleading concept has become part of established astronomical terminology. Thus when we speak of an 'early-type' star, or that the classification of one star is 'later' than that of another, the historical origin of these terms should be borne firmly in mind.

The H–R diagram also gives rise to yet another aspect of stellar classification. We have already come across stars on the main sequence, and the red giant and white dwarf stars. In fact several such subdivisions emerge on a detailed analysis of the way in which stars are distributed on the H–R diagram. This brings us to the topic of luminosity classification, in which stars are classified, not by their temperature, but by their luminosity. Stars on the main sequence are luminosity class V, while the giants are divided amongst the supergiants (luminosity class Ia and Ib), bright giants (luminosity class II), giants (luminosity class III) and sub-giants (luminosity class IV). Thus the complete classification of the Sun is G2V, whereas that of Rigel is B8Ia.

In general a star is interesting from our point of view when it is off the main sequence and this occurs during its formation (the pre-main sequence stage) and when the star has used up a substantial amount of its supply of fuel (the evolved state). The reason for this is two-fold. First, it is at these extreme stages of evolution that a star is likely to have an associated dusty envelope. Pre-main sequence stars often have debris (which is often dusty) left over from the star formation process; this debris is eventually dispersed by the star, usually by a stellar wind. [Note that in some cases–such as the Sun–some of this debris may remain to form planets.] At the other evolutionary extreme, highly evolved stars, especially those of later spectral type, often show enhancements of heavy elements and these stars are often extremely efficient producers of dust.

1.4.3 Stellar populations

In addition to the classification of stars by their temperature and luminosity, stars are also described by their membership of *Population I* or *Population II*. Population I stars generally lie in the plane of the Galaxy (see Section 1.5) and are responsible for the brightness of the spiral arms of our own and other galaxies; in particular, stars of spectral type O and B are Population I. Population II stars are generally found in the Galactic halo and, unlike Population I stars such as the Sun, tend to be underabundant in heavy elements. Indeed it is significant that Population II stars are generally found in regions that are dust-free. Observational evidence points to the fact that Population II stars were amongst the first to form in the early history of the Galaxy, whereas the formation of Population I stars is a process that is continuing in the spiral arms. However, as with spectral types, stars do not fit neatly into two distinct populations. Indeed in Chapter 9 the possibility will be discussed of a third population of stars which may have preceded the formation of the galaxies themselves during the early Universe.

1.4.4 Variable stars

Many stars in an early or advanced state of evolution are variable in their light output: the radiation we receive from them varies with time. Such stars are called variable stars and a plot of flux (or flux density, or magnitude, or any other equivalent quantity) against time is called a light curve. Obviously a light curve need not necessarily describe the variation of stellar *light* with time, but can describe the time-dependence of radiation in any part of the electromagnetic spectrum.

Stellar variability can either be intrinsic or extrinsic, in other words it can originate either in the star itself or in its immediate environment. In the former case the star may be pulsationally variable (e.g. Cepheid and RV Tauri variables) or may have surface inhomogeneities (e.g. starspots); another possibility is stellar flares. In others the variability is explosive, as it is in novae and supernovae. In the case of extrinsic variability the cause of flux variation lies outside the star itself and may arise, for example, as a result of the occultation of the star by a large cloud of circumstellar dust, or even by another star. The light curve of a variable star is, more often than not, a characteristic of the type of variability. For example, Cepheid variables, novae and various types of pre-main sequence stars (such as T Tauri stars) all have characteristic and easily identifiable light curves.

Variable stars are designated essentially according to the order in which they are discovered in a constellation. The sequence is R...Z, followed by RR...RZ, SS...SZ, TT...TZ, and so on up to ZZ. The sequence then continues with AA...AZ, BB...BZ and finally QQ...QZ. This peculiar system allows for a total of 335 variable stars in a constellation and if the number exceeds this (as it frequently does in constellations close to the Galactic plane) the sequence continues with V336, V337 etc. (V denoting variable). In each case the variable star designation is qualified by the genitive form of the Latin name of the constellation (e.g. RV Tauri, mentioned above).

This system does not accommodate supernovae, which have their own system of nomenclature. The first supernova discovered in any year is designated by the the letter 'A' following the year of discovery, the second 'B' and so on. Thus the third supernova to be discovered in 1979 is referred to as '1979C'.

1.5 The Galaxy and the interstellar medium

The space between the stars is not empty: in addition to the dust particles the properties of which are discussed in this book, there is also an interstellar gas whose properties have direct relevance to the study of cosmic dust grains.

The properties of this gas have been determined by observing its absorption effects at ultraviolet, optical and infrared wavelengths, and by its emission at radio and millimetre wavelengths. The gas consists mostly of hydrogen in atomic form but there is also an important molecular component, particularly CO. In addition, free electrons and the interstellar magnetic field–both of which have relevance to the study of interstellar dust grains–affect the propagation of electromagnetic radiation, mainly at radio wavelengths. In this section we briefly describe the properties of the interstellar gas and magnetic field.

The Galaxy has mass $\simeq 10^{11}\, M_\odot$ and consists of a flat circular disc (radius 15 kpc) of gas and Population I stars, and a central bulge of Population II stars (see Chapter 9). The disc is not uniform but consists of spiral arms traced out by hot, young stars. The thickness of the Galactic disc depends to a large extent on how it is measured. For example if the thickness of the HI layer is measured we find a value of 250 pc, whereas the thickness of the CO layer is about 120 pc; the thickness of the free electron layer is about 700 pc. The thickness of the dust layer is 200 pc.

It is possible to set an upper limit on the amount of matter in interstellar space by considering the motions of stars that move out of the Galactic disc. Such stars will experience the gravitational effects of all matter (stellar and non-stellar) in the plane of the Galaxy. The average density of matter in the plane of the Galactic disc can therefore be determined to be $10 \times 10^{-21}\,\mathrm{kg\,m^{-3}}$. Furthermore the average density of stars in the Galactic disc can also be determined, by undertaking star counts and integrating over the stellar mass distribution; the result is $4 \times 10^{-21}\,\mathrm{kg\,m^{-3}}$. However, such counts are likely to miss stars at the faint (low mass) end of the mass distribution so the above value for the average density of stars is a lower limit. We therefore have an upper limit, the Oort limit, on the amount of matter in interstellar space:

$$\rho_{\mathrm{Oort}} \leq 6 \times 10^{-21}\,\mathrm{kg\,m^{-3}}. \tag{1.21}$$

1.5.1 The interstellar gas

The interstellar gas consists mainly of neutral atomic hydrogen (HI), with number density $\sim 10^5$ H atoms $\mathrm{m^{-3}}$ and temperature about 100 K, although there are of course substantial variations around these typical values. Virtually all the atomic hydrogen in the interstellar medium is in the ground state; nevertheless this gas emits because the spins of the electron and proton render the ground state of the hydrogen atom degenerate, the energy difference corresponding to the well-known 1421 MHz transition. In addition there is a significant molecular component, observations of rotational transitions in the CO molecule in particular being invaluable in complementing the determination of the structure of the Galaxy using HI emission.

There are regions (e.g. molecular clouds) where the interstellar gas is substantially more dense than indicated above and the hydrogen is molecular, rather than atomic; in such clouds the number density of H_2 molecules can be anything up to 10^{10} molecules m^{-3}. The opacity in such clouds due to extinction by interstellar dust grains can be substantial and indeed, the dust plays a fundamental rôle in the chemistry of such clouds.

In the vicinity of a hot (spectral type O or B) star the hydrogen is ionized by the ultraviolet photons emitted by the star and, if the hydrogen is uniformly distributed, the star is surrounded by a spherical region in which the hydrogen is completely ionized; such regions are referred to as HII regions, or as Strömgren spheres, after the Swedish astrophysicist who first investigated such regions theoretically. One such region is shown in Fig. 1.2.

In addition to the atomic and molecular component, there is an ionic component and an electron component. The effect of the ionic (e.g. CaII, MgII) component is essentially similar to that of the atomic component and, like the latter, can be used to determine the abundances of the elements in the interstellar gas; as we shall see in Chapter 6, there are significant and substantial differences between the abundances of the elements in the interstellar gas and in other astronomical objects. The free electron population of the interstellar gas consists of two components: a relativistic component and a thermal component at an effective temperature of a few thousand degrees K. The latter has a number density of about $10^4 - 10^5$ electrons m^{-3} and gives rise to dispersion of low frequency electromagnetic radiation as it propagates through the interstellar medium.

1.5.2 The interstellar magnetic field

Evidence for the existence of an interstellar magnetic field comes from several quarters. First there is the Zeeman effect, which splits the HI 1421 MHz emission into two or three components, depending on the direction of the magnetic field. Second there is the Galactic synchrotron radiation, which arises as a result of the gyrations of interstellar relativistic electrons–mentioned above–in the field; since the radiation is linearly polarized in a direction that is perpendicular to the magnetic field vector these data provide information about both the magnitude and direction of the field. Third we can measure the Faraday rotation that arises because the free thermal electrons and magnetic field render the interstellar medium birefringent. All these observations suggest the existence of an interstellar magnetic field that is ordered along the spiral arms of the Galaxy and having magnitude $\sim$ a few times 10^{-10} T.

1.6 Gathering information about dust

Virtually all of what we know about the Universe in general, and the Dusty Universe in particular, has been deduced by detecting electromagnetic radiation, from the millimetre to X-rays. However, the Earth's atmosphere is opaque to most forms of electromagnetic radiation and so a large fraction of this radiation does not penetrate the atmosphere to reach the ground. Indeed, until it became possible to place observatories in orbit around the Earth, all our knowledge about the Universe came from optical, radio and a limited amount of infrared astronomy.

The transparency of the atmosphere to incoming radiation is determined (amongst other factors) by its chemical composition. At the longest radio wavelengths the atmosphere is opaque because the ionosphere reflects all radiation having wavelength longer than about 10 m back into space. At shorter radio and millimetre wavelengths water vapour and O_2 render the atmosphere opaque and molecular absorption in general determines the transparency in the infrared. However there are a number of narrow 'windows' in the infrared which can be exploited by carrying out observations at altitude. Indeed the effective wavelengths and widths of the broadband infrared filters described in Section 1.3.2 are, to a large extent, determined by the wavelengths and widths of the narrow infrared windows in the Earth's atmosphere. At shorter ($\sim$ ultraviolet) wavelengths, absorption by atomic and molecular nitrogen, atomic and molecular oxygen (including ozone) are important. In this section we describe the kind of information we get about dust in the Universe at various wavelengths. We begin with the longest wavelengths and work our way shortward.

1.6.1 Millimetre wavelengths

We shall see in Chapter 3 that spherical particles have an emissivity that is proportional to ν^α (α = a constant) at wavelengths that are long by comparison with the particle size. At typical interstellar grain temperatures a grain will emit radiation essentially on the Rayleigh-Jeans tail at millimetre wavelengths and at these wavelengths therefore we can expect the observed emission to have power-law dependence on frequency. Accordingly at millimetre wavelengths we get information about the emissivity of dust grains.

In addition, many molecules have rotational transitions at millimetre wavelengths and as molecules have an important rôle to play in the evolution of cosmic dust grains–both in circumstellar and interstellar environments–observations of molecules at millimetre wavelengths are of direct relevance to the investigation of dust grains.

1.6.2 Infrared

The infrared region may generally be taken to extend from 1 μm at the short wavelength end to overlap with the millimetre range at the long wavelength end. At infrared wavelengths we can also get information about the emissivity of dust grains. Most solids–cosmic or otherwise–evaporate if they are heated to temperatures in excess of about 2000 K. If we refer to Eq. (1.20) we see that a blackbody solid at temperature less than this value will emit radiation at wavelengths $\gtrsim 1.5\,\mu$m, i.e. in the infrared. Since real materials do not emit like blackbodies (the emission of dust grains is discussed in more detail in Chapters 6 and 7) the infrared flux as a function of wavelength will be determined, at least in part, by the emitting efficiency as a function of wavelength so infrared observations provide useful information about this parameter. In particular there are several important signatures in the emissivity curves of various grain types that are vital in identifying the composition of the emitting (or absorbing) grain. In the case of circumstellar dust, the observed infrared flux distribution is also determined by the range of grain temperatures and sizes present, so infrared observations may tell us about these quantities as well. Dust grains will also be rendered visible in the infrared by virtue of their scattering and extinction properties, so polarization studies and infrared imaging may give useful pointers to grain sizes.

As is the case with observations at shorter wavelengths, observations from Earth orbit have revolutionized our view of the infrared Universe. The first such observations were obtained by the Infrared Astronomical Satellite *IRAS*. *IRAS'* prime function was to conduct an all-sky survey in four wavebands, an objective that was superbly carried out during the satellite's 10 month operational lifetime. The observations consisted of broad-band photometry centred on the wavelengths 12, 25, 60 and 100 μm (designated bands 1–4 respectively); response functions of the *IRAS* detectors as a function of wavelength are shown in Fig. 1.8. Note how the 12 μm band overlaps with the 10 μm window accessible from the ground, so that a check is provided on the consistency of the fluxes measured by *IRAS* with those measured from the ground-based observatories. *IRAS* also had a low resolution ($\lambda/\Delta\lambda \sim 20 \rightarrow 60$) spectrometer, which provided spectra in the wavelength range 7–23 μm.

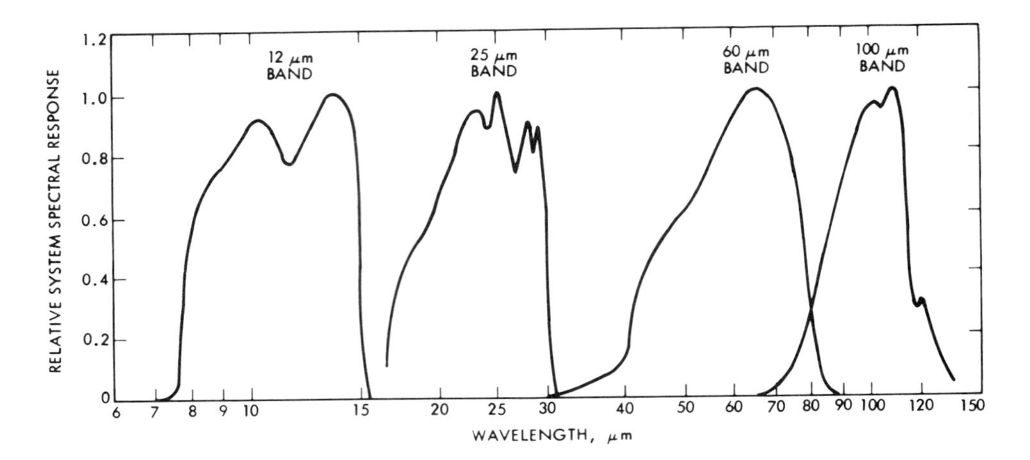

Figure 1.8: Wavelength response of *IRAS* bands. After G. Neugebauer et al., *Astrophysical Journal*, Vol. **278**, L1 (1984).

1.6.3 Optical

Historically the optical has provided a great deal of information about the extinction, scattering and polarization properties of cosmic dust. The presence of extinction was discovered by Herschel in the 18 th century during his extensive investigation into the distribution of the stars, although he actually attributed the apparent absence of stars in certain regions of the sky to 'holes' in the stellar distribution; examples of the effects of extinction are shown in Fig. 1.1, and in Fig. 1.9, which shows the Bok globule B338, taken in red and in blue light.

Investigations into the wavelength-dependence of interstellar extinction in the optical and near-infrared in the 1930's led to the first indication of the size of the obscuring particles; this was confirmed by later work on the dependence of polarization on wavelength. The wavelength-dependence of extinction is well illustrated in Fig. 1.9. The prime motive for most of the early work on the amount and wavelength-dependence of interstellar extinction was to enable observers to allow for the effect of the dust in observations of objects (like stars) in which they were really interested.

The existence of scattered light from interstellar dust grains was discovered by means of long exposure photographs of systems such as the Pleiades cluster of stars (see Fig. 1.10). Nebulosity such as that shown in Fig. 1.10 is referred to as *reflection nebulosity* and such nebulosity can arise in three ways. In the case of the Pleiades the cluster, in its motion through space,

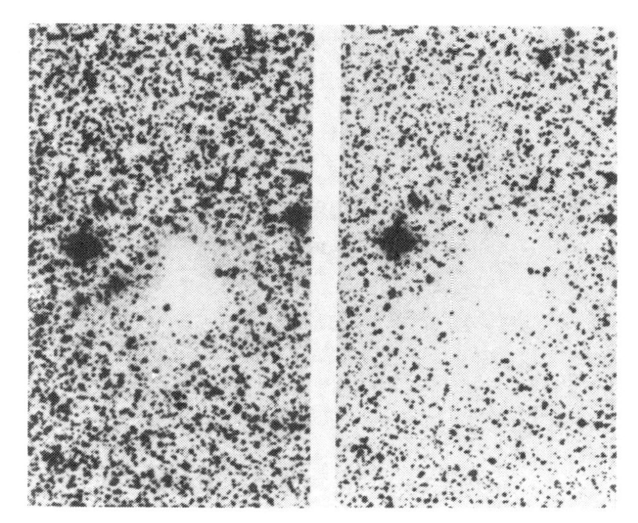

Figure 1.9: The Bok globule B338 in red (left) and blue (right) light; note the greater extinction in the blue. After B. T. Lynds, *Stars and Stellar Systems*, Vol. VII. University of Chicago Press. Eds G. Kuiper & B. Middlehurst (1975). Mount Wilson and Palomar Observatory photograph.

Figure 1.10: Scattering by dust grains in the vicinity of the Pleiades cluster. Photograph from the Hale Observatories.

happens to be passing through a cloud of interstellar dust grains; the grains are consequently illuminated by the light of the cluster stars. Secondly there is a diffuse 'glow' over much of the sky which arises as a result of the scattering of background starlight by dust particles in the interstellar medium. Thirdly individual stars may themselves be associated with dust particles, either of their own making or remaining from the process of star formation; such reflection nebulosity, associated with circumstellar dust, are characteristically very compact. The investigation of starlight reflected by dust particles provides information on the way in which individual particles scatter light. Maximum information is obtained only when the scattering geometry (star–grain–observer) is known; unfortunately in most cases this is not available as the three dimensions of space are projected onto a two-dimensional sky.

Optical spectroscopy can also provide indirect information about the composition of dust grains. Narrow absorption lines in stellar spectra can tell us something about the abundances of elements in interstellar space and thus provide constraints on the chemical composition of interstellar grains. In addition stars whose light is known to be affected by interstellar dust frequently have 'diffuse' absorption features in their spectra which have so far defied identification. There is much circumstantial evidence, however, that these features are associated with interstellar dust grains.

1.6.4 Ultraviolet

The extension of observational work to the ultraviolet in the past few years has proved invaluable in providing data about the extinction properties of interstellar and circumstellar dust.

The Earth's atmosphere becomes opaque to radiation with wavelengths less than about 3000 Å, and observations at shorter wavelengths have to be carried out from satellite observatories. Most of the early ultraviolet satellites (such as the Astronomical Netherlands Satellite *ANS*) provided broad-band (resolution $\lambda/\Delta\lambda \sim 2$) photometry, although *Copernicus*, which also carried out X-ray observations, provided spectroscopy at resolution $\lambda/\Delta\lambda \sim 500$.

The International Ultraviolet Explorer satellite *IUE* was launched in 1978 and provides for low ($\lambda/\Delta\lambda \sim 1000$) and high ($\lambda/\Delta\lambda \sim 10^4$) resolution spectroscopy in the wavelength range 1000–3200 Å and observations using this facility have provided a wealth of information about the extinction properties of interstellar and circumstellar dust over the first 15 years of operations.

Shortward of wavelength 912 Å the interstellar medium itself becomes opaque to radiation. This limit arises because a photon having wavelength less than this value (or, equivalently, energy > 13.6 eV) will, in view of the large cosmic abundance of hydrogen compared with other species in the interstellar medium, tend to ionize a hydrogen atom. Atomic hydrogen has a number density of about 10^5 atoms m^{-3} in the general interstellar medium, and the ionization cross-section (at the ionization threshold) is about 3×10^{-21} m^2. A photon having energy > 13.6 eV (a *Lyman continuum* photon) therefore travels only about 3×10^{15} m (about 0.1 pc) before it ionizes a hydrogen atom and is effectively lost. Since the nearest star is at a distance of 1.3 pc it may seem that few Lyman continuum photons are capable of reaching the terrestrial observer from all but the nearest astronomical objects.

1.6.5 X-ray

From what was said in the previous subsection about the opacity of the interstellar medium, it may seem surprising that X-ray astronomy is possible at all. However, the ionization cross-section of hydrogen declines steeply with decreasing frequency (roughly as ν^3) with the result that, at soft X-ray wavelengths (shortward of about 100Å) the interstellar medium once more becomes transparent to radiation. It is this fortunate circumstance that allows us to carry out X-ray astronomy.

The earliest attempts at detecting X-radiation from celestial objects involved crude detectors (such as photographic emulsions) mounted on rockets which carried the detector above the Earth's atmosphere for literally a few minutes' observation (the Earth's atmosphere being, of course, opaque to X-radiation). Further significant developments had to await the launch of satellite observatories, such as *Uhuru* and the *Ariel* series, which carried out surveys of the X-ray sky in the photon energy range 100 eV to 2 keV (wavelength range 0.1–100 Å).

It may seem surprising that X-ray wavelengths have any relevance to the study of cosmic dust. After all, if we refer to Eq. (1.20) we see that thermal emission at X-ray wavelengths (30 Å say) will arise in material at temperatures of $\sim 10^6$ K and no solid–cosmic dust grains included–can exist at such high temperatures. However, it was predicted in the 1960's that interstellar dust grains would scatter X-radiation with the result that suitable X-ray sources would appear to possess diffuse haloes of a few arcminutes' extent (see Chapter 6). Thus while facilities (such as *Uhuru*) provided the capability of detecting and measuring the fluxes of X-ray sources in the late 1960's and 1970's, it was only with the advent of facilities having an imaging capability (such as the *Einstein* and *EXOSAT* satellites) that the interest in the scattering of X-rays by interstellar dust revived.

Furthermore, dust grains immersed in a hot gas may be heated by the impacts of ions; this process can be significant if the gas is at temperatures exceeding $\sim 10^6$ K, at which temperatures it will emit at X-ray wavelengths. The effect of grain heating by ion impact can be significant in supernova remnants, many of which are strong X-ray emitters.

Problems

1.1. Show how an equation like (1.14) follows from the Wien approximation to the Planck function.

1.2. Show that Eq. (1.17) conforms with the definition of magnitude differences and flux ratios.

1.3. Determine the radius (in $R_\odot$) of a star having (i) $L_{\rm bol} = 10^4\,L_\odot$, effective temperature $T_{\rm eff} = 3500$ K; (ii) $L_{\rm bol} = 10^{-3}\,L_\odot$, effective temperature $T_{\rm eff} = 10^4$ K.

Reading

For a glossary of astronomical terms see

[B] *Glossary of Astronomy and Astrophysics*, J. Hopkins, University of Chicago Press (1980).

The book by Kitchin provides a good introduction, at the undergraduate level, to observational techniques:

[C] *Astrophysical Techniques*, C. R. Kitchin, Adam Hilger (1988).

The following book is an imaginative and novel discussion of astronomical observations in the various bands of the electromagnetic spectrum:

[A] *Cosmic Landscape*, M. Rowan-Robinson, Oxford University Press (1980).

An account of stellar properties, evolution etc. may be found in the texts listed above under 'Introductory reading', and also in

[C] *Introduction to Stellar Astrophysics*, in three volumes, E. Böhm-Vitense, Cambridge University Press (1989–1992).

2

From astronomy to physics

2.1 Introduction

Unlike physics or chemistry, both of which are experimental sciences, astronomy is an observational science: the observer can take no part in, and can not affect, the phenomena he observes. If we are to understand the physics of astronomical phenomena, we must be able to relate the intensity of measured radiation (as measured in an observatory) to the physical laws that govern the emission of that radiation, i.e. to whatever physical process is going on in the source under observation.

The emission of radiation is ultimately governed by the laws of quantum physics (e.g. via Einstein coefficients, or in the case of dust particles, the distribution of phonon modes in a solid). Once the radiation is emitted, it is likely to be absorbed and scattered on its way to the observer and again, the laws that govern these processes are determined by the basic laws of physics. In this chapter we look at the way in which the transition is made from astronomy to physics by showing how the measured intensity of the radiation depends on essentially physical phenomena.

2.2 The transfer of radiation

2.2.1 The equation of transfer

As radiation crosses the space between a source and the observer its intensity can be affected by interaction with matter. We will consider in more detail the interaction of radiation with matter (specifically dust grains) in Chapters 3 and 4 but for the moment we just take it for granted that there are radiation losses and gains as the radiation propagates towards the observer,

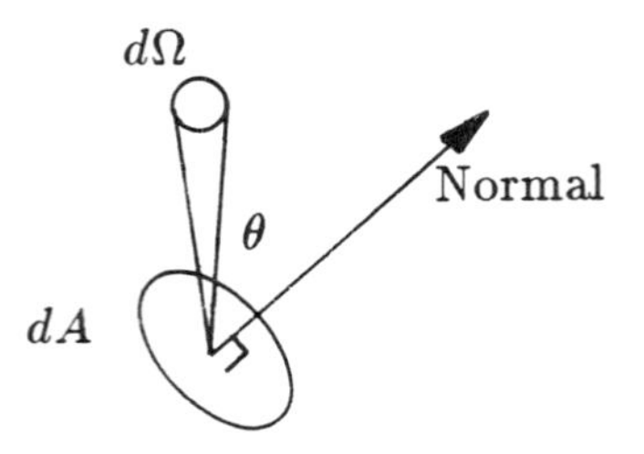

Figure 2.1: Definition of the intensity of radiation.

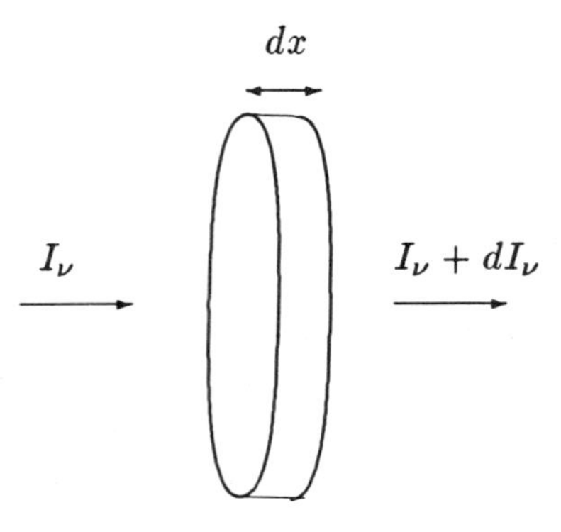

Figure 2.2: Extinction in an infinitesimal pill-box.

without worrying too much about the details of *how* radiation is produced and lost.

The *intensity* of radiation is formally defined as follows. Suppose we have an amount of energy dE_ν, in the frequency range $\nu \rightarrow \nu + d\nu$, in the form of radiation; the radiation is confined to a solid angle $d\Omega$, and crosses an element of area dA in time dt, at an angle θ to the normal to dA (see Fig. 2.1). The intensity I_ν is defined such that

$$dE_\nu = I_\nu \cos\theta d\nu dA d\Omega dt. \tag{2.1}$$

In a medium that scatters, absorbs and emits radiation, I_ν will vary from point to point in the radiation field, and with direction. If I_ν is independent of direction the radiation field is said to be isotropic, and if it is independent of position it is said to be homogeneous.

Suppose we have a ray which enters an infinitesimally thin pill-box, of thickness dx (see Fig. 2.2). Let the intensity of radiation on one side of the pill-box be I_ν and, as a result of the radiation losses resulting from the interaction of radiation with matter within the pill-box, the intensity at the other side is $I_\nu + dI_\nu$. The *extinction coefficient* κ_ν is defined by

$$dI_\nu = -\kappa_\nu I_\nu dx. \tag{2.2}$$

In general, the process of extinction involves *absorption*, in which radiation is physically removed from the beam and *scattering*, in which the radiation suffers a change of direction of propagation; in either case radiation is lost from the point of view of the observer. If there are n absorbers or scatterers per unit volume, each of cross-section σ_ν, then $\kappa_\nu = n\sigma_\nu$. We can also define the *optical depth*

$$\tau_\nu = \int_{\text{pathlength}} \kappa_\nu dx. \tag{2.3}$$

The optical depth may be visualized as follows. Suppose we have a medium (like a fog) that absorbs and scatters light. We know from experience that we can only see a certain distance into the fog: the 'visibility' may be poor. Knowing the mechanism by which radiation is scattered or absorbed [i.e. knowing κ_ν in Eq. (2.3)], the visibility is given, to a good approximation, by that distance for which $\tau_\nu \simeq 1$. This distance is, of course, likely to be wavelength-dependent because of the wavelength-dependence of κ_ν.

Eq. (2.2) gives the reduction in intensity following the loss of radiation in the pill-box. Matter in the pill-box may also emit radiation and the total change in intensity on traversing the length of the pill-box is

$$dI_\nu = \underbrace{-\kappa_\nu I_\nu dx}_{\text{extinction}} + \underbrace{\epsilon_\nu dx}_{\text{emission}} . \tag{2.4}$$

The units of the *emission coefficient* ϵ_ν are energy emitted per unit time per unit volume per unit frequency interval into unit solid angle. It is also helpful to define the *source function* $\mathcal{S}_\nu$

$$\mathcal{S}_\nu = \frac{\epsilon_\nu}{\kappa_\nu}. \tag{2.5}$$

We can now rewrite Eq. (2.4) in the form

$$\frac{dI_\nu}{dx} = -\kappa_\nu I_\nu + \epsilon_\nu \tag{2.6}$$

or, by using Eqs. (2.3) and (2.5), as

$$\frac{dI_\nu}{d\tau_\nu} = -I_\nu + \mathcal{S}_\nu. \tag{2.7}$$

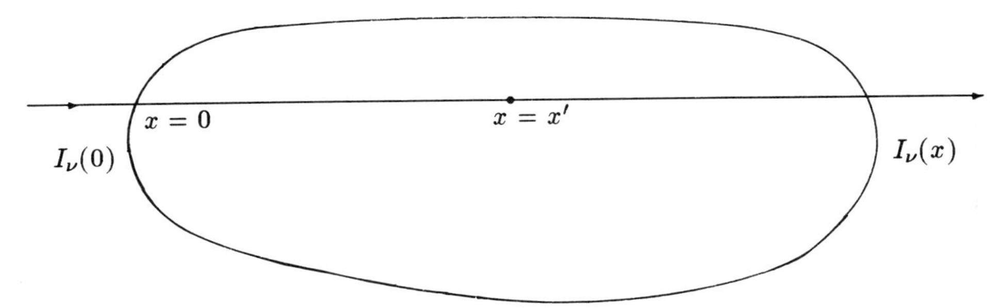

Figure 2.3: Calculating the intensity of radiation.

Eqs. (2.6) and (2.7) are different versions of the *equation of transfer* which, by solving for I_ν, allows us to relate the observations (represented by I_ν) to the physical processes going on in the source (represented by ϵ_ν and κ_ν). Although the equation of transfer looks harmless enough its solution, even for the simplest of cases, is generally very complicated and usually has to be done numerically.

2.2.2 Solving the equation of transfer

We can solve Eq. (2.7) to give I_ν explicitly in terms of the other quantities. Suppose we have radiation of intensity $I_\nu(0)$ incident on one side of a source of total extent x in the line of sight (see Fig. 2.3). We wish to solve Eq. (2.7) to determine the intensity $I_\nu(x)$ emerging at the other side of the source. If the source function $\mathcal{S}_\nu$ is independent of I_ν then the equation is a standard linear differential equation. This is solved by multiplying throughout by an integrating factor $\exp[\tau_\nu]$; if we do this and rearrange we get

$$d[I_\nu \exp(\tau_\nu)] = \exp(\tau_\nu)\mathcal{S}_\nu d\tau_\nu, \tag{2.8}$$

where the optical depth τ_ν in Eq. (2.8) is

$$\tau_\nu = \tau_\nu(0, x') = \int_0^{x'} \kappa_\nu dx, \tag{2.9}$$

x' being some intermediate value of x along the line of sight (see Fig. 2.3). In Eq. (2.9) $\tau_\nu(0, x')$ is simply the optical depth between $x = 0$ and $x = x'$. We can now integrate Eq. (2.8) between 0 and x, i.e. along the entire path-length through the source:

$$I_\nu(x) \exp[\tau_\nu(0, x)] - I_\nu(0) = \int_0^x \exp[\tau_\nu(0, x')]\mathcal{S}_\nu d\tau_\nu. \tag{2.10}$$

We now multiply Eq. (2.10) throughout by $\exp[-\tau_\nu(0,x)]$ and take the $I_\nu(0)$ term over to the right hand side:

$$I_\nu(x) = I_\nu(0)e^{-\tau_\nu(0,x)} + \int_0^x \exp[-\tau_\nu(x',x)]\mathcal{S}_\nu d\tau_\nu, \qquad (2.11)$$

where we have used the fact that $\tau_\nu(0,x) = \tau_\nu(0,x') + \tau_\nu(x',x)$.

Eq. (2.11) is one of the most basic and fundamental equations in astrophysics, because it relates physical processes (represented by ϵ_ν and κ_ν), which are determined by the laws of physics, to the radiation intensity, which is what the observer detects. It provides the 'connecting link'–the interface–between physics and astronomy.

Although we seem to have 'solved' the equation of transfer what we have really done is just to rearrange it: Eq. (2.11) is really only another form of Eq. (2.7). Our 'solution' only applies if $\mathcal{S}_\nu$ is independent of I_ν and this is not necessarily the case. Nonetheless Eq. (2.11) can be used to calculate the intensity of radiation in cases where there are no complications such as multiple scattering of radiation, optically thick media etc.

Let us examine the solution (2.11) and see what the physical significance is of the two terms on the right hand side. The first term

$$I_\nu(0)e^{-\tau_\nu(0,x)}$$

represents radiation incident on the farside of the source $I_\nu(0)$, attenuated by the factor $e^{-\tau_\nu(0,x)}$ by the effects of extinction by material within the source. Indeed this result could have been obtained directly by integrating Eq. (2.2), i.e. in the absence of any *emission* of radiation. The $e^{-\tau_\nu}$ term in Eq. (2.11) represents the probability that a ray starting at $x = 0$ reaches x.

If we refer back to Eqs. (2.3) and (2.5), we can see that the term $\mathcal{S}_\nu d\tau_\nu$ in the integrand of Eq. (2.11) can be written as

$$\mathcal{S}_\nu d\tau_\nu = \epsilon_\nu dx;$$

this term therefore represents the emission by material around the position $x = x'$ within the source itself. However, the emitted radiation is itself attenuated by an amount $e^{-\tau_\nu(x',x)}$ by the material within the source. The second term

$$\int_0^x \exp[-\tau_\nu(x',x)]\mathcal{S}_\nu d\tau_\nu$$

therefore represents the contribution to the intensity by the source itself, summed over the line of sight through the source.

In some applications of course there is no background radiation, in which case $I_\nu(0) = 0$; in others we know that the source does not emit at the

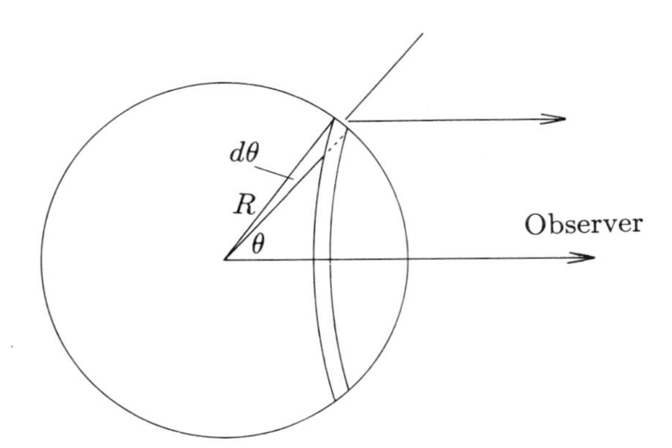

Figure 2.4: Calculating the flux from a spherical source.

frequencies of interest, so that $\mathcal{S}_\nu = 0$. Obviously every application of the radiative transfer equation has to be tailored to the problem at hand.

The solution Eq. (2.11) gives the intensity of radiation emerging through an element of area dA at the end of a line of sight through the source. Frequently the nature of the detecting system used is such that the source is not resolved and appears essentially as a point source. Under these circumstances we can integrate the intensity over the solid angle subtended by the source [cf. Eq. (1.7)] to get the flux density. In the simple case of a spherical source of radius R (see Fig. 2.4) the solid angle is

$$d\Omega = \frac{2\pi R \sin\theta R d\theta}{D^2}$$

and we have

$$S_\nu = \int_0^{\pi/2} I_\nu \frac{2\pi R^2}{D^2} \sin\theta \cos\theta d\theta. \tag{2.12}$$

Note that Eq. (2.12) is consistent with Eq. (1.8) if I_ν is independent of θ.

2.3 Equivalent width

One frequently finds in the spectra of stars absorption lines which do not arise in the stellar atmosphere but which arise as a result of absorption by material in the space between the star and the observer. Stellar absorption lines can always be distinguished by the fact that they are Doppler shifted by an amount that corresponds to the motion of the star relative to the observer; they are also rotationally broadened and have a characteristic profile.

As has already been mentioned, absorption lines that arise in the interstellar medium come in two varieties. The one arises in the interstellar gas,

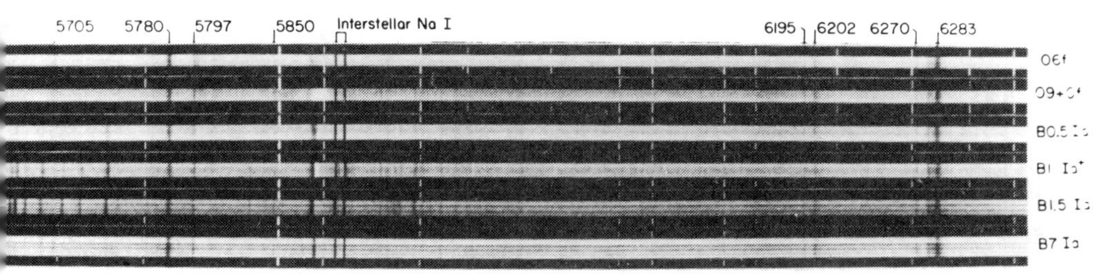

Figure 2.5: Interstellar sodium D lines in absorption. Also present are several diffuse interstellar bands at the wavelengths indicated. Emission lines above and below each spectrum are included to calibrate the wavelength scale. Photograph from the Lick Observatory and the Royal Astronomical Society.

and these lines can generally be distinguished from stellar lines by virtue of (i) their sharpness and (ii) the fact that the material in which they originate does not share the velocity of the star. The latter property is well illustrated in Fig. 2.5, in which the spectra of six hot stars have been aligned at the interstellar sodium D lines. If this figure is viewed at glancing incidence the fact that the *stellar* absorption lines (e.g. the one close to the interstellar sodium lines) are at slightly differing wavelengths can clearly be seen. These interstellar lines are extremely narrow (see Fig. 2.5)–so much so that their precise profile is often difficult to measure, even at high spectral resolution. The second variety of interstellar absorption lines are more enigmatic and although there is every indication that they arise in interstellar space there is, in general, no universally accepted identification. It has long been suspected that these lines originate in, or are in some way associated with, interstellar dust grains. A number of these lines are present in Fig. 2.5 and we return to these features in Chapter 6.

In either case it is convenient to describe the strength of the lines in terms of their *equivalent width*, which is defined as follows. We consider the absorption line shaded in Fig. 2.6. The spectrum around the line is assumed to be adjusted so that the continuum is flat, or it is supposed that the continuum level does not change appreciably across the width of the line. The equivalent width is defined as that width of continuum the area of which is equal to the area contained by the line profile (see Fig. 2.6). The area under the continuum is, from the definition of equivalent width, $S^0_\nu W_\nu$, where the equivalent width W_ν is measured in frequency units and S^0_ν is the height of the continuum on either side of the line. The area contained by the line profile is $\int [S^0_\nu - S_\nu] d\nu$, where the integral is carried out over the line

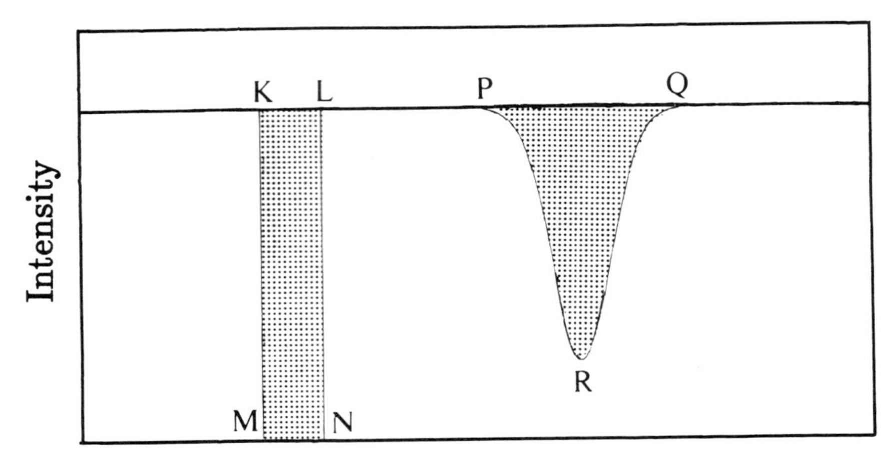

Figure 2.6: Definition of the equivalent width of an absorption line. The areas PQR (between the line profile and the stellar continuum) and $KLMN$ are equal. The equivalent width is KL.

profile. Thus

$$W_\nu = \int \left[1 - \frac{S_\nu}{S_\nu^0}\right] d\nu \tag{2.13}$$

where, for later convenience, we have taken S_ν^0 inside the integral sign (note that it is independent of frequency). It is more customary to measure equivalent width in wavelength units. Converting Eq. (2.13) using Eq. (1.3), we have

$$W_\lambda = \frac{\lambda^2}{c} \int \left[1 - \frac{S_\nu}{S_\nu^0}\right] d\nu. \tag{2.14}$$

Clearly with this definition of equivalent width $W > 0$ for an absorption line and < 0 for an emission line.

Problems

2.1. Show that, to first order, the observed flux density S_ν from an optically thin ($\tau_\nu \ll 1$), homogeneous source of volume V at distance D from the observer is $S_\nu = V\epsilon_\nu/D^2$, where ϵ_ν is the emission coefficient.

2.2. A uniform slab of emitting material has emission coefficient $\epsilon_\nu = A\nu^{-\alpha}$, where A and α are constants; a plasma has absorption coefficient $\kappa_\nu = B\nu^{-2}$ (B constant). Determine the form of the resulting intensity if the material is (i) seen through a uniform slab of plasma and (ii) well-mixed with the plasma. Assume that the plasma is non-emitting and that the emitting material does not absorb or scatter its own radiation.

3

The interaction of a grain with radiation

3.1 General ideas

It is the way in which a grain interacts with radiation that determines a number of its most important observational characteristics, and we should devote some time to the discussion of this first. This is not as trivial a problem as it might appear at first sight because, as we shall see in Chapter 6, the dimensions of interstellar grains are comparable with a wavelength of ultraviolet-visible radiation. The investigation of the properties of dust grains can of course proceed on two fronts, namely a 'theoretical' treatment, in which one tries to understand the basic physics of what is going on; and a 'laboratory' study, in which the properties of interest are measured directly in the laboratory. We consider the former aspect in this chapter and we begin by recalling some ideas from solid state physics and considering the nature of this interaction.

3.2 Phonon modes in solids

3.2.1 Acoustic modes

The simplest lattice of all consists of a linear chain of identical atoms, the distance between adjacent atoms being denoted by a. Suppose that the nth atom is displaced from its equilibrium position by a small amount δ_{n}, where the displacement δ_{n} is parallel with the length of the chain. As a result of the displacement of the nth atom the neighbouring atoms will also be displaced, by amounts $\delta_{\mathrm{n-1}}, \delta_{\mathrm{n+1}}$ etc. (see Fig. 3.1). The nth atom is

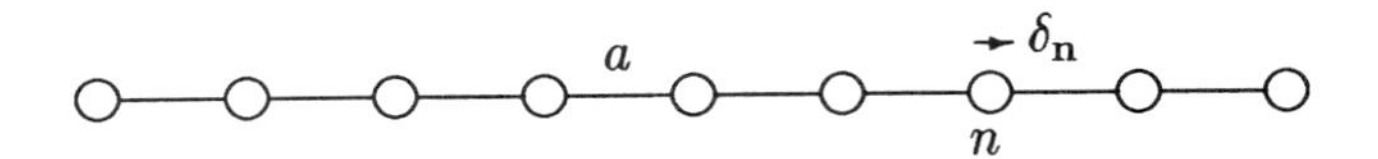

Figure 3.1: A linear chain of identical atoms. In equilibrium the atoms are separated by a distance a but the nth atom is displaced by δ_n in a direction that is parallel with the chain.

therefore displaced by an amount $\delta_n - \delta_{n-1}$ relative to the $(n-1)^{th}$ atom, and by an amount $\delta_n - \delta_{n+1}$ relative to the $(n+1)^{th}$ atom etc. The nth atom experiences a restoring force F that, like Hooke's law, depends linearly on the displacement from its equilibrium position. Since there is an atom on either side of the nth atom the *net* restoring force is

$$F = -\eta[(\delta_n - \delta_{n-1}) - (\delta_{n+1} - \delta_n)], \tag{3.1}$$

where η is the force constant. According to Newton's second law, this restoring force is also given by

$$F = m\frac{d^2\delta_n}{dt^2},$$

where m is the mass of an atom. Thus

$$\frac{d^2\delta_n}{dt^2} = \frac{\eta}{m}[\delta_{n-1} + \delta_{n+1} - 2\delta_n]. \tag{3.2}$$

Note that this approach does not take into account the restoring forces arising from next-nearest and further neighbours, but this does not qualitatively affect the conclusion that follows. We look for a solution of Eq. (3.2) in wave form:

$$\delta_n = \delta_0 \exp[i(kna - \omega t)]. \tag{3.3}$$

Here ω is the (angular) frequency of the wave and k the wavenumber, defined in terms of the wavelength λ in the usual way as $k = 2\pi/\lambda$; δ_0 is the amplitude of the oscillation. Substituting δ_n from Eq. (3.3) in Eq. (3.2) we get

$$\omega^2 = \frac{2\eta}{m}(1 - \cos ka) \tag{3.4}$$

$$= \left(\frac{4\eta}{m}\right) \sin^2 \frac{ka}{2} \tag{3.5}$$

so that

$$\omega = \pm 2 \left(\frac{\eta}{m}\right)^{1/2} \sin\left(\frac{ka}{2}\right). \tag{3.6}$$

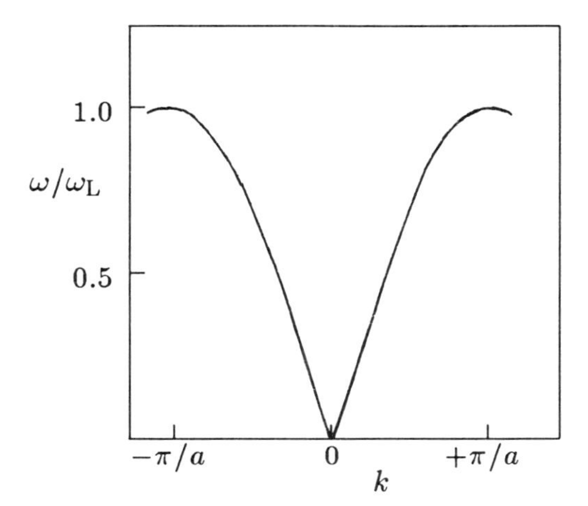

Figure 3.2: Dispersion relation for longitudinal vibrations in a linear lattice.

Eq. (3.6) gives the dispersion relation that describes the allowed vibrational modes in the lattice, the positive and negative signs denoting waves travelling in the directions of increasing and decreasing n respectively. Note that, since the displacement δ_n is parallel with the length of the chain the wave is longitudinal; this is referred to as an acoustic mode. From Eq. (3.6) we see that modes having frequencies greater than a critical frequency $\omega_L = 2(\eta/m)^{1/2}$ can not propagate through the lattice because values of $\omega \geq 2(\eta/m)^{1/2}$ are forbidden. The dispersion relation (3.6) is illustrated in Fig. 3.2 and is plotted in the usual way in the form ω against k.

More generally individual atoms in the chain will also be able to vibrate transversely as well as longitudinally. In general therefore we can expect three separate dispersion relations, one corresponding to the longitudinal vibrations already discussed and two corresponding to transverse vibrations in two mutually perpendicular directions. However, in the simplest cases the two transverse modes will not be distinguishable: they will be degenerate.

3.2.2 Optical modes

There are obviously more possibilities if, instead of a chain of identical atoms, we have a chain consisting of alternate atoms of different kinds A and B, having masses m_A and m_B respectively with $m_A > m_B$ (see Fig. 3.3). The equilibrium separation between the atoms is again d. The nth atom, of type A, undergoes a small displacement δ_n from its equilibrium position. The

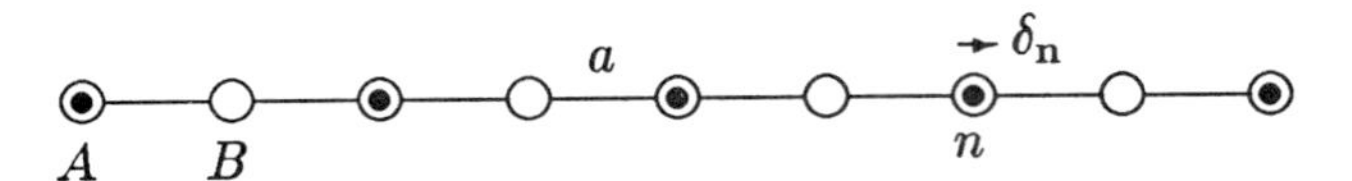

Figure 3.3: A linear chain consisting of alternate atoms of different kinds. As in Fig. 3.1 the atoms are separated by a distance a but the nth atom is displaced by δ_n in a direction that is parallel with the chain.

nearest-neighbour atoms, both of which are of type B in a chain of alternate atoms, are displaced by amounts δ_{n-1} and δ_{n+1} respectively. The *next-nearest* neighbours to the nth atom are of type A, and suffer displacements δ_{n-2} and δ_{n+2} respectively, and so on all along the chain.

Since we have two different atom types in the chain we now require two equations for the restoring force, one for each atom:

$$F_A = -\eta_1[(\delta_{n-1} - \delta_n) - (\delta_n - \delta_{n+1})], \tag{3.7}$$
$$F_B = -\eta_2[(\delta_n - \delta_{n+1}) - (\delta_{n+1} - \delta_{n+2})]. \tag{3.8}$$

Note that, as we have two different atoms the restoring force η_1 on atom A will be different from that on atom B, η_2; however for simplicity we shall assume $\eta_1 = \eta_2 = \eta$. Again we can write

$$F_A = m_A \frac{d^2\delta_n}{dt^2} \tag{3.9}$$
$$F_B = m_B \frac{d^2\delta_{n+1}}{dt^2} \tag{3.10}$$

and we look for solutions of the form

$$\delta_n = \delta_A \exp[i(kna - \omega t)] \tag{3.11}$$
$$\delta_{n+1} = \delta_B \exp[i(k\{n+1\}a - \omega t)]; \tag{3.12}$$

δ_A and δ_B now describe the amplitudes of the atoms A and B respectively.

Proceeding as before we find that the dispersion relation is now of the form

$$\omega^2 = \eta\left(\frac{1}{m_A} + \frac{1}{m_B}\right) \pm \eta\left\{\left[\frac{1}{m_A} + \frac{1}{m_B}\right]^2 - \frac{4\sin^2(ka)}{m_A m_B}\right\}^{1/2}, \tag{3.13}$$

which opens up more possibilities than Eq. (3.4); the dispersion relation, again in the form of a plot of ω against k, is shown in Fig. 3.4. Taking the

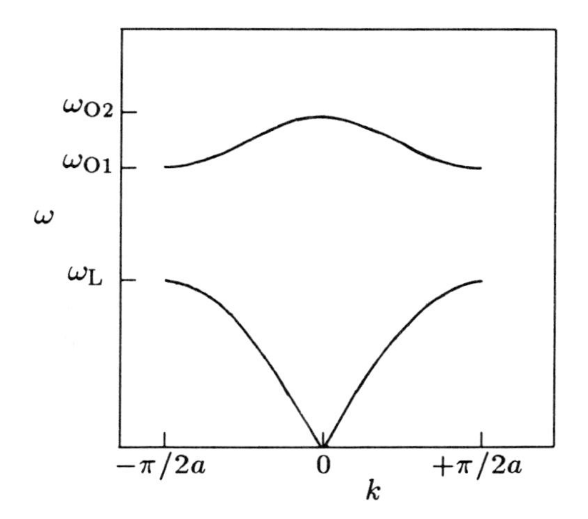

Figure 3.4: Dispersion relation for optical vibrations in a linear lattice.

minus sign in Eq. (3.13) gives us the longitudinal (acoustic) modes, already encountered (see Fig. 3.2). There is again a maximum frequency:

$$\omega_L = \left(\frac{2\eta}{m_A}\right)^{1/2},$$

corresponding to the maximum allowed frequency in the simple monatomic chain and, as in the case of the simple monatomic chain, the vibrations are longitudinal. However in this case we can also have ***transverse*** acoustic vibrations, the atoms A and B oscillating in phase.

Taking the positive sign in Eq. (3.13) leads to the optical mode, so called because this mode is easily excited by radiation of the appropriate frequency in ionic solids. In the optical mode there are both minimum

$$\omega_{O1} = \left(\frac{2\eta}{m_B}\right)^{1/2}$$

and maximum

$$\omega_{O2} = \left[2\eta\left(\frac{1}{m_A} + \frac{1}{m_B}\right)\right]^{1/2}$$

frequencies. In this case the atoms A and B oscillate out of phase and again, transverse vibrations are possible in which the atoms vibrate out of phase. Similar conclusions apply if the simple monatomic chain of Fig. 3.1 is replaced by two atoms (identical or otherwise) occupying the same lattice site.

We note that ω is complex for frequencies between ω_L and ω_{O1}, irrespective of the value of k; this range of ω defines the *forbidden band*, in which lattice vibrations are heavily damped.

The phonon velocity in solids $3.3\,\mathrm{km\,s^{-1}}$ for carbon to about $6.5\,\mathrm{km\,s^{-1}}$ for silicates and silicon carbide. [Note that for an anisotropic material like graphite, however, the sound velocity will depend on the direction of propagation.] A photon of frequency ν has of course energy and momentum $h\nu$ and $h\nu/c$ respectively and the process of converting photons to phonons must obviously conserve momentum and energy. However the fact that $V_{\mathbf{phonon}} \ll c$ means that the conversion from photon to phonon modes is not straightforward. We note from Fig. 3.4, however, that optical mode phonons with frequency close to ω_{O2} may have momentum $\hbar k$ close to zero while having considerable energy ($\hbar\omega_{O2}$). This is where the conversion from photon to phonon modes can be made provided that a means of effecting this conversion is available. With $\eta \sim 10^3\,\mathrm{kg\,m^{-2}}$ and taking typical values of m_A and m_B (e.g. $m_A = 12\,\mathrm{amu}$ for C, $m_B = 28\,\mathrm{amu}$ for Si) we find that the required frequency is in the infrared part of the spectrum; thus an infrared photon can, in principle, be converted to an optical mode phonon. To further understand how the conversion takes place we next recall the electrical properties of solids.

3.3 The electrical properties of solids

The polarizability of a solid has its origins in three distinct microscopic phenomena, namely (i) the contribution of the dipole moments of polar molecules in the solid (polarization due to dipole orientation); (ii) the relative movement of positive and negative ions in the solid (ionic polarization) and (iii) the displacement of atomic electrons relative to the atomic nuclei (electronic polarization). The first two will not necessarily contribute to the polarizability of every solid, the third being the only one that is present in all dielectric materials.

In a static electric field, all three of the above phenomena may, where appropriate, contribute to the polarizability of the solid. In a field that varies sufficiently slowly with time, each of the above contributions varies in phase with the external field. Under these circumstances the dielectric function is purely a real number. At higher frequencies the contribution due to any polar molecules is unable to keep up with the varying electric field–there arises a phase difference between the incident wave and the molecules and the dielectric becomes 'lossy': the electromagnetic wave at these frequencies (typically about 10^{10} Hz) is damped. Consequently the real part of the

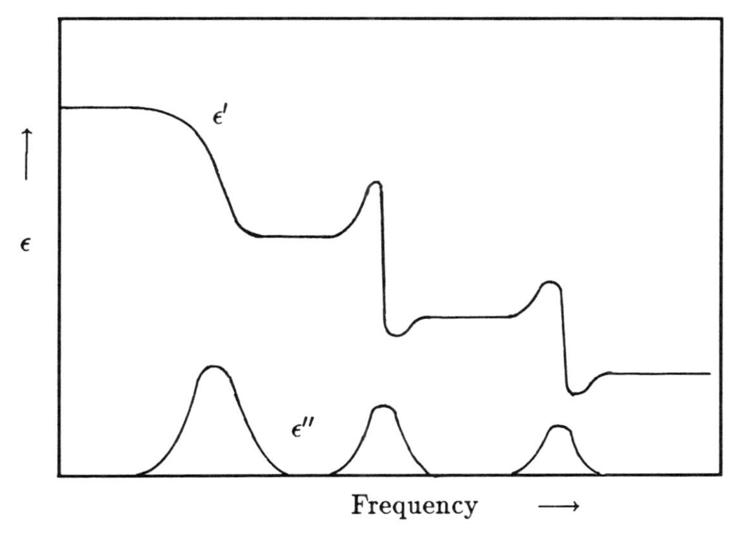

Figure 3.5: Schematic dependence of complex dielectric function on frequency.

dielectric function decreases as the contribution of polar molecules drops out and the imaginary part becomes non-zero[1]. A similar effect occurs (generally in the infrared, $\nu \sim 10^{13}$ Hz) in ionic solids, when the frequency of the incident electromagnetic wave is such that the dipoles formed by positive and negative ions fail to keep up with the oscillations of the electromagnetic wave. Finally, the electronic contribution to the dielectric function fails to stay in phase at high ($\sim 10^{15}$ Hz) frequencies, beyond which the dielectric function falls to unity for most solids. The dependence of the real and imaginary parts of the dielectric function on frequency is illustrated schematically in Fig. 3.5.

For the reasons outlined above the real and imaginary parts of the dielectric function are not independent; they are related by the *Kramers-Kronig* relations,

$$\epsilon'(\omega) = 1 + \frac{1}{\pi}\int_{-\infty}^{\infty} \frac{\epsilon''(x)}{x-\omega} dx \tag{3.14}$$

$$\epsilon''(\omega) = -\frac{1}{\pi}\int_{-\infty}^{\infty} \frac{\epsilon'(x)-1}{x-\omega} dx \tag{3.15}$$

where $\omega = 2\pi\nu$ is the circular frequency. Consequently it is generally sufficient to know either ϵ' or ϵ'' as a function of frequency and the other can

[1]Note that the imaginary part of the dielectric constant must be positive otherwise the electromagnetic wave is amplified rather than damped; see Section 3.4.2 below.

be derived from Eq. (3.14) provided the known quantity is available over a sufficiently large range of frequencies.

Positive and negative ions in an ionic material will be accelerated in opposite directions by an electromagnetic wave, i.e. optical mode phonons will be excited. For an electromagnetic wave whose E−vector is described by $E_0 \exp[-i\omega t]$, the transverse displacements of the ions, δ', may be determined from the equation for a damped harmonic motion

$$\ddot{\delta}' + \gamma\dot{\delta}' + \omega_{O2}^2\delta' = -\frac{eE_0 \exp[-i\omega t]}{m}.$$

The solution is

$$\delta' = -\frac{eE}{m}\frac{1}{\omega_{O2}^2 - \omega^2 - i\gamma\omega},$$

where the constant γ has the effect of damping the electromagnetic wave and provides a mechanism whereby the energy in the wave can be converted to optical phonon modes and thence to other phonon modes.

In an electrically conducting material the effect of an electric field (such as that associated with an electromagnetic wave) is of course the flow of current, of density

$$\mathbf{J} = \sigma_E\, \mathbf{E}, \tag{3.16}$$

where σ_E is the electrical conductivity of the material. The conduction electrons are scattered by imperfections in the solid, such as impurity atoms, dislocations etc. and inelastic scattering results in the creation of phonons: the effect is again that energy in the electromagnetic wave is converted to phonon modes. In this case we expect that the absorbing efficiency of the grain is related to σ_E and this is indeed the case.

3.4 Absorption, extinction and scattering

We shall see in the following chapters that the dust grains in interstellar and circumstellar space have dimensions $\sim 0.1\ \mu$m. A detailed investigation of the interaction of electromagnetic radiation with small particles leads to the result that the cross-sectional area for the interaction of a spherical grain of radius a with radiation is not given by the 'obvious' result πa^2 but by $Q\pi a^2$, where the quantity Q (which is ~ 1) depends on the nature of the grain (its refractive index, size etc.) and the wavelength of the radiation involved. Furthermore, we have to specify whether the process we are interested in is extinction, scattering or absorption. Scattering of radiation results in a change in the direction of propagation without any change in wavelength (although in general the *efficiency* with which scattering takes place may well be wavelength-dependent). Absorption results in the physical removal of radiation from the propagating ray and leads to the heating of the absorbing particle. However from the point of view of a distant observer who merely detects a loss of intensity it is not, in general, possible to know whether absorption, or scattering, or both, have been instrumental in removing radiation from the line of sight between the source and the observer: what the observer measures is extinction, the combined effect of absorption and scattering. For each of these processes there is a corresponding Q-factor; thus, for example, the cross-section for the absorption of radiation by a spherical grain of radius a is given by $Q_{\mathrm{abs}}\pi a^2$.

The cross-section for the extinction of radiation by a spherical grain of radius a is given by $\sigma_{\mathrm{ext}} = Q_{\mathrm{ext}}\pi a^2$; there are similar definitions of σ_{sca} and σ_{abs} for scattering and absorption respectively. Since extinction is the sum of both scattering and absorption, we can write

$$\sigma_{\mathrm{ext}} = \sigma_{\mathrm{abs}} + \sigma_{\mathrm{sca}} \tag{3.17}$$

so that, for a spherical grain,

$$Q_{\mathrm{ext}} = Q_{\mathrm{abs}} + Q_{\mathrm{sca}}. \tag{3.18}$$

It is also useful to define the *albedo* ϖ; this is the ratio of scattering to extinction efficiencies:

$$\varpi = \frac{Q_{\mathrm{sca}}}{Q_{\mathrm{ext}}}. \tag{3.19}$$

In addition to the optical efficiencies Q there is another quantity of interest that comes out of the same analysis, namely the *scattering phase function*. We consider a plane-parallel wavefront incident on a spherical grain which scatters the radiation, and we wish to know the angular distribution of the scattered radiation: how does the intensity of the scattered radiation depend on the angle at which the radiation is scattered? This question is answered by the scattering phase function $S(\theta)$, which is defined as follows (see Fig. 3.6). If the incident intensity is I_0, the intensity of the radiation scattered at angle θ to the incident direction is given by $I_0S(\theta)$; clearly in this case the scattering diagram has axial symmetry about the direction of the incident radiation.

However it is often not necessary to have detailed knowledge of $S(\theta)$: an average value of the scattering angle is all that is required. This average is often denoted by specifying a value of $\cos\theta$ which has been suitably averaged over all directions. This average, denoted by $\langle\cos\theta\rangle$ or by g, is sufficient for some purposes. Note that, if $g = \langle\cos\theta\rangle$ is close to unity, then the radiation is largely scattered in the forward direction, whereas if g is close to zero, the

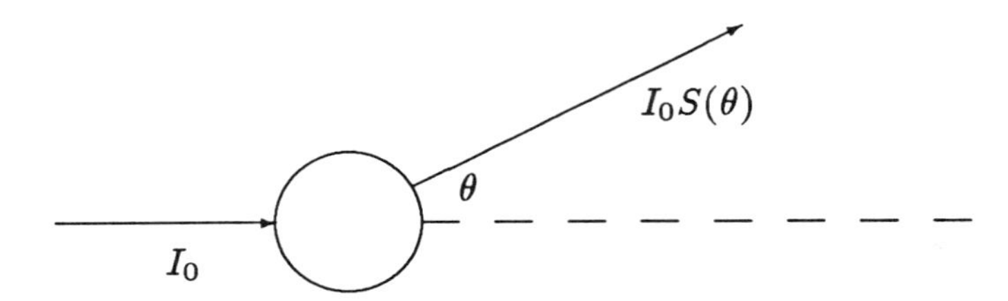

Figure 3.6: Definition of scattering phase function.

scattering is isotropic, i.e. the radiation is scattered equally in all directions. We have

$$g = \langle \cos\theta \rangle = \frac{\int_0^\pi S(\theta)\cos\theta\sin\theta d\theta}{\int_0^\pi S(\theta)\sin\theta d\theta}$$

In Fig. 3.6, scattering in directions concentrated around $\theta \simeq 0°$ is referred to as *forward-scattering*, while scattering in directions around $\theta \simeq 180°$ is referred to as *back-scattering*. Only rarely do we have a case in which the scattering angle is known. This can arise when, for example, a star is known to lie in front of a dust cloud which scatters its light; in this case we see light that has been back-scattered and in a case like this it is useful to define a Q–factor for back-scattering Q_{bk}.

3.4.1 Radiation pressure

As is well known, the electric and magnetic components of an electromagnetic wave oscillate at right angles to the direction of propagation (i.e. the wave is transverse) and in addition, the electric and magnetic vectors are mutually perpendicular. Most of the observable effects of an electromagnetic wave is due to the electric component. However, it is essentially the magnetic component that gives rise to the phenomenon of radiation pressure. The electric component of the wave gives rise to a current $\mathbf{J} = \sigma_{\mathrm{E}}\mathbf{E}$ in grain material of electrical conductivity σ_{E}. The *magnetic* component of the electromagnetic wave $\mathbf{B}$ then exerts a force $\mathbf{J} \times \mathbf{B}dxdydz$ on a current element, in the direction of propagation of the wave. For example, in the case of an electromagnetic wave propagating in the x–direction, with electric and magnetic components E_{y} and B_{z} respectively, $J_{\mathrm{y}} = \sigma_{\mathrm{E}}E_{\mathrm{y}}$ and the force exerted on a current element is $J_{\mathrm{y}}B_{\mathrm{z}}dxdydz$; the force per unit area (i.e. the pressure) is thus $dp_{\mathrm{x}} = J_{\mathrm{y}}B_{\mathrm{z}}dx$.

We now derive the efficiency factor for radiation pressure. Radiation carries momentum as well as energy. Recall that momentum is a vector quantity and so its direction, as well as its magnitude, must be specified. Now the direction of the vector associated with the momentum of radiation is simply the direction of propagation. Again suppose that we have radiation incident on a small spherical particle of radius a. As we have already seen, some of this radiation is absorbed by the grain, the amount absorbed [cf. Eq. (2.2)] being proportional to the absorption coefficient, which in turn is proportional to the absorption cross-section $Q_{\rm abs}\pi a^2$. This absorbed component is forever lost to the incident ray and plays no further part in the discussion. The situation as regards the radiation scattered by the particle is different. Radiation that is scattered at angle θ to the incident direction still has a component of momentum in this direction, so not all the incident momentum is lost by scattering. Now the total amount lost by scattering is (for spherical particles) proportional to the scattering cross-section $Q_{\rm sca}\pi a^2$. But from what has already been said an amount proportional to $Q_{\rm sca}\langle\cos\theta\rangle\pi a^2$ is resupplied by the scattering process so the *net* amount lost by scattering is simply proportional to $[Q_{\rm sca} - Q_{\rm sca}\langle\cos\theta\rangle]\pi a^2$. Thus the total amount of incident momentum removed from the incident beam is proportional to

$$[Q_{\rm abs} + Q_{\rm sca} - Q_{\rm sca}\langle\cos\theta\rangle]\pi a^2 = [Q_{\rm ext} - Q_{\rm sca}\langle\cos\theta\rangle]\pi a^2, \tag{3.20}$$

where we have used the relation (3.18). The quantity in square brackets in Eq. (3.20) defines $Q_{\rm pr}$, the efficiency factor for radiation pressure, as

$$Q_{\rm pr} = Q_{\rm ext} - Q_{\rm sca}\langle\cos\theta\rangle. \tag{3.21}$$

3.4.2 The variation of the Q-factors

The calculation of the various Q factors, of $S(\theta)$ and of g is extremely complicated but in general we would expect these quantities to depend on the geometry of the grain (e.g. whether spherical, cylindrical or whatever), on the orientation of the grain relative to the incident ray, and also on the refractive index of the grain material. The problem of investigating the way in which electromagnetic radiation interacts with a homogeneous spherical object was first solved by G. Mie in 1908 and, independently, by P. Debye in 1909; the solution was later extended by Güttler in 1952 to include homogeneous spherical grains coated with material having different optical properties. Solutions are also available for homogeneous ellipsoidal and cylindrical particles. However, the problem is generally referred to under a general umbrella of *Mie theory.*

For many materials of interest as far as cosmic dust is concerned, the refractive index of the grain material depends on wavelength, and is a complex quantity. It is the imaginary part of the refractive index that determines the extent to which radiation is absorbed. To see this, write the complex refractive index m as

$$m = n' - in'', \tag{3.22}$$

where n' is the real part of the refractive index and n'' the imaginary part. Assume that the incident electromagnetic wave can be expressed as a sinusoidal vibration, of amplitude A and circular frequency ω, propagating along the positive x-direction:

$$A = A_0 \exp[i(\omega t - kx)]. \tag{3.23}$$

In a medium of refractive index m the velocity of light is given by c/m and so the amplitude of the ray will be given by

$$A = A_0 \exp[i(\omega t - kxn')] \exp[-kn''x]. \tag{3.24}$$

Note that the intensity I of the wave, which is proportional to A^2, decreases exponentially with a decay constant of $2kn''$. This shows that the absorption coefficient of the material is directly related to the imaginary part of the refractive index n''. The complex refractive index is related to the complex dielectric function $\epsilon = \epsilon' + i\epsilon''$ by

$$\epsilon' = n'^2 - n''^2 \tag{3.25}$$

$$\epsilon'' = 2n'n''. \tag{3.26}$$

The determination of the Q factors for grain materials of various types requires knowledge of the values of n' and n'', which have been measured in the laboratory for a number of possible grain materials. The results of calculations for spherical grains with various refractive indices are given in Fig. 3.7, in which the various Q's are plotted against the quantity $x = 2\pi a/\lambda$. Although the details differ, there are a number of common features in the curves displayed. For example, we see that Q_{ext} and Q_{sca} both increase with increasing x when x is very small ($\ll 1$); in fact both Q_{sca} and Q_{abs} are proportional to x^α for particles whose dimensions are small by comparison with the wavelength, where α is generally in the range $1 \rightarrow 2$. Thus for example $Q_{\mathrm{abs}} \propto \lambda^{-\alpha} \propto \nu^\alpha$. Also, both Q_{abs} and Q_{sca} approach unity as x becomes very large–i.e. for objects large by comparison with incident wavelength. In other words, the cross-section for scattering or absorption is simply the geometrical cross-section of the grain; this is the sort of result that we might expect intuitively. Note that the curves level off when x is about unity, i.e. when the radius of the sphere $\gtrsim \lambda/2\pi$. We also see a series of maxima and minima, which arise as a result of interference between incident, scattered and transmitted rays.

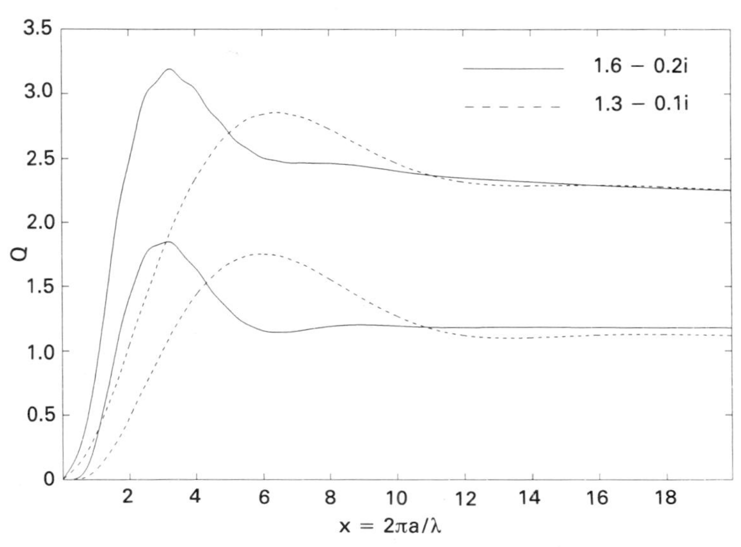

Figure 3.7: Dependence of optical efficiencies Q_{ext} and Q_{sca} on $x = 2\pi a/\lambda$. Refractive indices as indicated; Q_{ext} is the upper curve in each case.

However the behaviour of Q_{ext} is surprising. Recall that the Q-factors give the ratio of the cross-section for interaction with radiation to the geometrical cross-section, and we might reasonably expect these factors to be unity in the limit of very large particles. Although the increase of Q_{ext} with increasing x is similar to that of Q_{abs} and Q_{sca} for large x we see that Q_{ext} approaches two: a large grain intercepts (by absorbing and scattering) *twice* as much radiation as its geometrical cross-section would lead us to expect. The explanation for this strange result outlined by H. C. van de Hulst in his classic book *Light Scattering by Small Particles* is superb and cannot be bettered:

> This noteworthy paradox, that a large particle removes from the incident beam exactly twice the amount of light it can intercept, has attracted attention from various sides. Its paradoxical character is removed if we recall the exact assumptions that have been made in its derivation. They are (a) that all scattered light, including that at small angles, is counted as removed from the beam, and (b) that the observation is made at a very great distance, i.e. far beyond the zone where a shadow can be distinguished. A flower pot in a window prevents only the sunlight falling on it from entering a room, and not twice this amount, but a meteorite of the same size somewhere in interstellar space

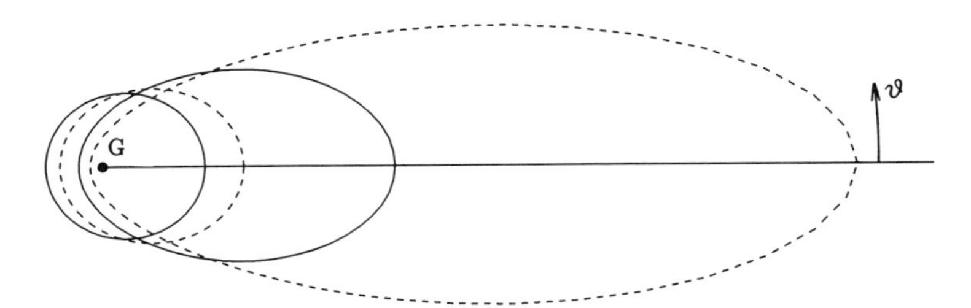

Figure 3.8: The Heyney-Greenstein scattering function for scattering by a grain at G; radiation is incident from the left. The polar diagrams are (in order of increasing ellipticity) for $g = 0.1, 0.2, 0.4, 0.6$.

> between a star and one of our big telescopes will screen twice this light. If many such rocks filled interstellar space, the light scattered by reflection and refraction would be seen as a faint luminosity all over the sky, and the diffracted light would show as narrow halos of light around each star. If the brightness of the star without this halo is measured, we have to use $Q = 2$. If, however, the method of measurement included the entire diffraction halo with the star's image (and this is certainly true if the telescope objective is smaller than the piece of rock), then it would be legitimate to use $Q = 1$.

The calculation of $S(\theta)$ and of g $(= \langle\cos\theta\rangle)$ proceeds along similar lines to the calculation of the Q-factors. However, there is an empirical formula that satisfactorily describes the behaviour of $S(\theta)$ for the case of forward scattering by large ($\gtrsim 0.05\,\mu$m) particles, namely the *Heyney-Greenstein* function:

$$S(\theta) = [1 - g^2][1 + g^2 - 2g\cos\theta]^{-3/2}. \tag{3.27}$$

Note that, if g is close to unity, $S(\theta)$ is negligible unless θ is near zero, in other words the radiation is strongly forward scattered. However, if g is close to zero, S is about unity in all directions, i.e. the scattering is isotropic (see Fig. 3.8). However, the Heyney-Greenstein scattering function does not describe the scattering by small ($\lesssim 0.05\,\mu$m) particles, nor back-scattering by larger particles. Evidently this formula is consistent with the comments already made.

Finally we note that, although there is no simple formula that gives the Q factors explicitly, there are approximation formulae that are valid for spherical particles when x is small. These are given by

$$Q_{\rm sca} = \frac{8}{3}x^4\Re\left\{\left(\frac{m^2-1}{m^2+2}\right)^2\right\} \qquad (x \ll 1) \qquad (3.28)$$

and

$$Q_{\rm abs} = -4x\Im\left(\frac{m^2-1}{m^2+2}\right) \qquad (x \ll 1) \qquad (3.29)$$

where the symbols $\Re$ and $\Im$ denote that the real and imaginary parts of the expressions should be taken. Readers familiar with Rayleigh scattering will recognize the characteristic λ^{-4} dependence of the scattering efficiency in Eq. (3.28).

However, it is not always useful to calculate $Q_{\rm ext}$ etc. for 'hypothetical' materials, with unrealistically constant values of n' and n''. As laboratory studies of likely grain materials demonstrate, real materials have refractive indices with real and imaginary parts that vary significantly with wavelength. Indeed, for this reason, the λ^{-4} dependence predicted by Eq. (3.28) may or may not result for scattering by small particles if the refractive index is wavelength-dependent. This dependence of m on wavelength is reflected in the behaviour of the Q–factors and we describe here some results of calculating $Q_{\rm sca}$ and $Q_{\rm ext}$ for materials that are likely to be of interest in the context of this book, namely carbon and silicate.

The dependence of $Q_{\rm ext}$ and $Q_{\rm sca}$ on inverse wavelength for spherical silicate grains of radius $0.1\,\mu$m is shown in Fig. 3.9. Note the peaks in the absorption efficiency at 9.7 μm ($\lambda^{-1} \simeq 0.1\,\mu$m^{-1}) and at 18 μm ($\lambda^{-1} \simeq 0.06\,\mu$m^{-1}), which arise as a result of the stretching of the Si–O and bending of the O–Si–O bonds respectively and are a characteristic of all silicates (such as olivine, forsterite etc.). We will encounter these features again when we discuss circumstellar dust in Chapter 7.

The dependence of $Q_{\rm abs}$ on inverse wavelength for carbon is shown in Fig. 3.10, for spherical grains of radius $0.01\,\mu$m and $0.1\,\mu$m. For the larger grains, the general similarity with the curves of Fig. 3.7 is evident. In the case of the smaller grains, however, we see that there is a broad peak, centred around 2100 Å ($\lambda^{-1} \simeq 4.75\,\mu$m^{-1}); this will be of interest when we come to discuss the interstellar extinction curve in Chapter 6.

3.4.3 Polarization

The scattering of radiation has a useful observational consequence. Consider the situation in Fig. 3.11, in which a ray of linearly polarized light from a

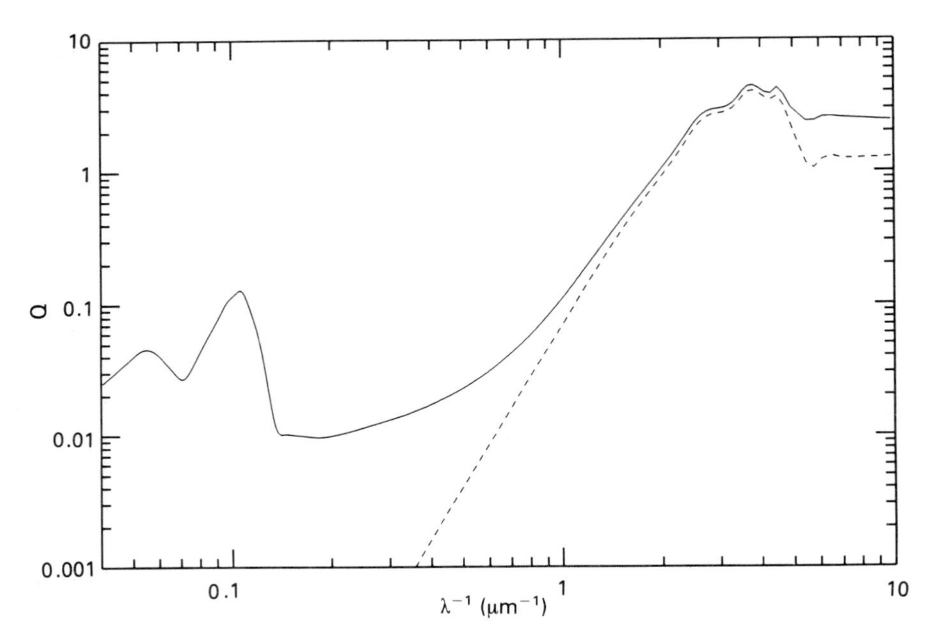

Figure 3.9: Dependence of Q_{ext} (solid curve) and Q_{sca} (dashed curve) on λ^{-1} for spherical silicate grains of 0.1 μm radius.

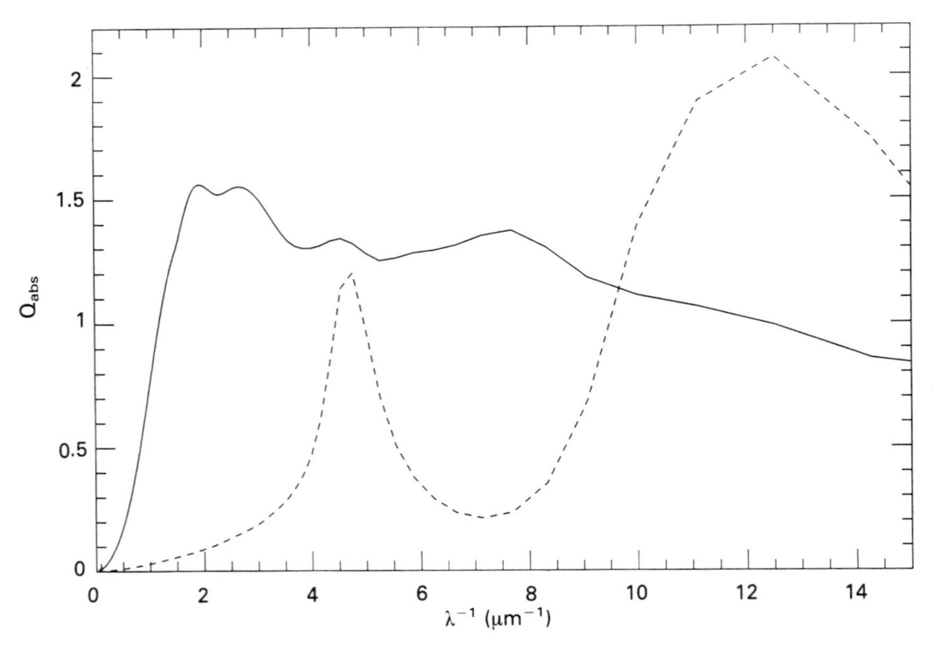

Figure 3.10: Dependence of Q_{abs} on λ^{-1} for spherical carbon grains of 0.1 μm (solid curve) and 0.01 μm (dashed curve) radius.

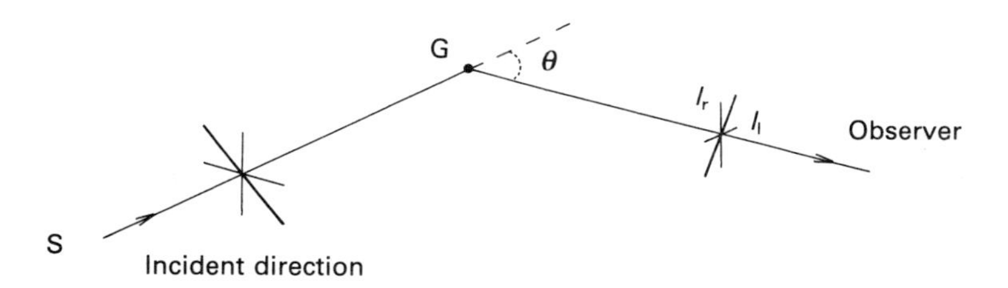

Figure 3.11: Scattering of linearly polarized radiation by a grain at G. The scattering plane contains the incident and scattered rays. The electric vectors of the incident and scattered radiation (thick lines) may be resolved with components (thin lines) parallel and perpendicular to the scattering plane.

source S, is scattered by a grain G; the scattering angle is θ. We can see that the incident and scattered rays define a plane, referred to as the scattering plane. The electric vectors of both incident and scattered rays may be resolved, with components parallel with, and perpendicular to, the scattering plane; it is customary to denote these components by subscripts l and r respectively. The component of the scattered intensity I_l may be close to zero for θ close to $\pi/2$ so that the scattered light in this direction is linearly polarized with electric vector oscillating perpendicular to the scattering plane; for forward- and back-scattering I_l and I_r are comparable, resulting in little polarization. At intermediate values of θ the degree of polarization

$$p = \frac{I_r - I_l}{I_r + I_l}$$

and the scattered ray is again polarized perpendicular to the scattering plane, i.e. to the plane defined by SGO. This result also applies even if the incident ray is unpolarized (as would be the case with starlight). This property will prove to be useful when we look at reflection nebulae in Chapter 6.

3.4.4 Elongated grains

The polarization of initially unpolarized radiation can also occur if the extinction of radiation having different senses of polarization is different. This might arise, for example, if the absorbing material (e.g. a dust grain) is elongated, or if it has impurities that are not uniformly distributed.

Obviously, for a homogeneous spherical grain, the orientation of the grain relative to the incoming radiation is irrelevant. However for any other grain geometry (e.g. cylindrical) the orientation of the grain with respect to the incident radiation must be specified and the values of the various Q-factors depend on orientation. This dependence of Q on orientation is important in considering the wavelength-dependence of interstellar polarization, to be discussed in Chapter 6.

For definiteness, we look at the extinction of a plane-polarized electromagnetic wave by a cylindrical grain and consider two cases. First, the case in which the electric vector of the wave oscillates in a plane that contains the major axis of the cylinder and second, in which the plane of oscillation is perpendicular to the cylinder axis. In the latter case it is the magnetic field associated with the wave that oscillates along the cylinder axis, so we denote the extinction efficiency for this case by $[Q_{\rm ext}]_{\rm H}$; the extinction efficiency for the former case we denote by $[Q_{\rm ext}]_{\rm E}$.

Qualitatively we expect these two Q-factors to be different for the following reason. Consider the case of a conducting cylindrical grain and an electromagnetic wave whose electric field oscillates in a plane that contains the axis of the cylinder. In such a case an electric current will flow along the axis of the cylinder and, as already discussed, collisions between the conduction electrons and imperfections in the grain will dissipate the electrons' energy, i.e. the grain will be heated. The net result is that an electromagnetic wave having this particular state of polarization will be damped. However for the case in which the *magnetic* vector oscillates in a plane that contains the cylinder axis the effect on any conduction electrons in the grain material is minimal and an electromagnetic wave that has its electric vector *perpendicular* to the cylinder axis proceeds relatively unimpeded. The net result is that radiation that was unpolarized before encountering a population of elongated grain emerges with a net polarization provided that the long axes of the grains are aligned.

The calculation of $[Q_{\rm ext}]_{\rm E}$ and of $[Q_{\rm ext}]_{\rm H}$ is again not straightforward but proceeds along similar lines to the calculation of $Q_{\rm ext}$. There is an added complication here of course in that the orientation of the cylinder axis with respect to the direction of propagation will affect the determination of both $[Q_{\rm ext}]_{\rm E}$ and $[Q_{\rm ext}]_{\rm H}$. However for normal incidence and infinitely long cylinders the following approximation formulae apply [cf. Eqs. (3.28), (3.29)]:

$$[Q_{\rm ext}]_{\rm E} \simeq -\frac{\pi x}{2}\Im\left(m^2-1\right) \tag{3.30}$$

$$[Q_{\rm ext}]_{\rm H} \simeq -\pi x \Im\left(\frac{m^2-1}{m^2+1}\right), \tag{3.31}$$

where again the symbol $\Im$ indicates that the imaginary part of the expressions should be taken. We shall see in Chapter 6 how the dependence of polarization on wavelength gives us a direct measure of the size of interstellar grains.

Problems

3.1. Determine (i) the scattering efficiency, (ii) the absorption efficiency, (iii) the extinction efficiency at wavelength 1 μm for spherical grains of radius 0.01 μm and the following refractive indices: $m = 1.6$, $m = 1.6 - 0.5i$.

3.2. A uniform spherical dust cloud, of radius R, contains n identical dust grains per unit volume; the grains have radius a, absorption efficiency Q_{abs}, and are in equilibrium at the same temperature T_{d}. There is no background radiation. Assuming that extinction within the cloud occurs by absorption only, use Eqs (2.11) and (2.12) to show that the flux density measured by an observer at distance D from the cloud is

$$S_\nu = \frac{2\pi R^2 B_\nu(T_{\mathrm{d}})}{\tau_0^2 D^2} \left(\frac{\tau_0^2}{2} - 1 + (\tau_0 + 1) \exp[-\tau_0] \right),$$

where $\tau_0 = 2\pi a^2 Q_{\mathrm{abs}} R n$ is the optical depth through the centre of the cloud.

3.3. What is the form of S_ν (Problem 3.2) in the limits (i) $\tau_0 \gg 1$ and (ii) $\tau_0 \ll 1$?

Reading

The following book is an excellent introduction to solid state physics:

[C] *Solid State Physics*, J. S. Blakemore, Cambridge University Press (1985).

The following has been the standard reference work on the interaction of radiation with small particles for nearly half a century:

[D] *Light Scattering by Small particles*, H. C. van de Hulst, J. Wiley & Sons (1957).

The paper by Draine & Lee gives a comprehensive discussion of the optical properties of graphite and silicates, while the paper by Draine is a useful compilation of complex refractive indices for both materials.

[D] B. T. Draine & H. M. Lee, *Astrophysical Journal*, Vol. **285**, 89 (1984).

[D] B. T. Draine, *Astrophysical Journal Supplement*, Vol. **57**, 587 (1985).

4

Properties of dust particles

4.1 Introduction

In this chapter we discuss the various factors that determine the physical properties of dust grains and, although the discussion is in an astronomical context, the principles apply equally well to the properties of a dust grain wherever it is situated.

From an observational point of view an important grain property is its temperature. The fact that a grain is 'hot' will result in its emitting radiation, and it is this emission that enables us to detect grains directly, by observing at infrared and millimetre wavelengths. Grains can also be observed directly by virtue of the fact that they scatter radiation, while the presence of an opaque dust cloud will be made evident by the fact that there is an obvious 'hole' in the distribution of stars (cf. Fig. 1.9). Generally speaking, then, grains are rendered observable by the way in which they interact with radiation.

However there are a number of other physical properties which, although not directly observable, can have considerable effect on the way in which dust grains evolve. Thus for example, a grain may acquire an electric charge, which may protect it from (or render it vulnerable to) collision and erosion by ions. Also, the efficiency with which atoms or molecules 'stick' to a grain surface will affect the rate at which a grain can grow. While grain erosion and growth are not in themselves observable, their effect can indirectly be seen in a number of astronomical environments.

4.2 Grain charge

There are several circumstances under which a dust grain might acquire electric charge. First, a grain immersed in a plasma will be bombarded by electrons and ions. Some of the electrons will 'stick' to the grain surface, so that the grain becomes negatively charged. The ions will not necessarily stick to the grain surface, but many of them will combine with surface electrons and 'lift' them off, thereby reducing the grain's negative charge. Second, a grain located close to a hot star which has a large flux of ultraviolet photons will eject electrons by virtue of the photoelectric effect. However not all grain materials will be subject to the photoelectric effect–only those which are electrical conductors. In this section we look at the various processes that contribute to the electric charge on a grain.

4.2.1 The photoelectric effect

Generally speaking electrons are strongly bound in a solid but those of highest energy may overcome the surface potential barrier. This may be achieved in a variety of ways. For example, if certain materials are irradiated by short-wavelength radiation they emit electrons by virtue of the photoelectric effect. As is well known, for a given material, this process only occurs for photons having energy greater than a specific value $h\nu_1$; this threshold is known as the *work function* $\phi = h\nu_1$. The energy E of the ejected photoelectron depends on the energy $h\nu$ of the incident photon via $E = h\nu - h\nu_1$.

If we have a spherical grain of radius a, its cross-section for the absorption of radiation is given by $\pi a^2 Q_{\rm abs}$ (see Section 3.4). However not every absorbed photon gives rise to a photoelectron. The photoelectric process has a *yield* $Y_{\rm p}$, which is the number of ejected electrons per incident photon; we know, again from laboratory work, that $Y_{\rm p}$ depends on photon energy, on the material being irradiated and also on the nature of the surface.

If the grain is situated a distance r from a source of radiation of luminosity L_ν, the flux density at frequency ν at the grain is

$$\frac{L_\nu}{4\pi r^2}.$$

However we are not interested in the flux density but in the number of photons being intercepted by the grain (recall how $Y_{\rm p}$ is defined in terms of *numbers* of electrons and photons); the number of photons in the frequency range $\nu \rightarrow \nu + d\nu$ arriving at the grain is

$$\frac{L_\nu}{4\pi r^2 h\nu} d\nu.$$

Each of these photons gives rise to $Y_{\rm p}$ photoelectrons so the total rate at which photoelectrons are emitted is

$$\int_{\phi_{\rm eff}}^{\infty} \frac{L_\nu}{4\pi r^2 h\nu} Y_{\rm p} \pi a^2 Q_{\rm abs} d\nu. \tag{4.1}$$

We should note that, if the grain is positively charged, it becomes more difficult for the electrons to escape from the grain: the coulomb attraction between grain and electron effectively increases the photoelectric work function of a grain of radius a by an amount given by $Ze^2/4\pi\epsilon_0 a$, where Ze is the positive charge on the grain. The result is that the threshold energy for the photoelectric effect is unchanged if the grain is negatively charged and is given by $h\nu_1 + Ze^2/4\pi\epsilon_0 a$ if the grain is positively charged. We can therefore define an effective work function

$$\begin{aligned} \phi_{\rm eff} &= h\nu_1 & (Ze \leq 0) \\ &= h\nu_1 + \frac{Ze^2}{4\pi\epsilon_0 a} & (Ze > 0). \end{aligned}$$

This effective work function is included in Eq. (4.1). To give some idea of how quickly a grain acquires positive charge we consider the simple situation of a 0.1 μm grain located a distance $r = 10^{16}$ m from a B5 star of luminosity $800 L_\odot$ and effective temperature 15 500 K. For a grain material having work function 2 eV, photoelectric yield 10^{-2} and $Q_{\rm abs} = 1$, we find from Eq. (4.1) that an electron is ejected from the surface of the grain every 10 seconds or so.

However as the positive charge on the grain increases as a result of the photoelectric effect, it will become increasingly attractive to any free electrons; in fact, as we shall see below, the cross-section for the collision of an electron with a positively charged grain exceeds the geometrical cross-section of the grain. In reality therefore equilibrium will eventually be attained when photoelectric emission is balanced by the attachment of electrons to the grain surface.

4.2.2 Thermionic effect

The distribution of electron energies in an electrical conductor at temperature $T_{\rm d}$ is determined by the Fermi-Dirac distribution. Electrons may have the energy necessary to overcome the surface potential by virtue of the fact that they occupy the highest available energy states in the solid; as with the photoelectric effect, if the electron energy exceeds the work function for the material under consideration it can escape from the surface of the solid. This process is referred to as *thermionic emission* and the flow of electrons is called the *thermionic current*. Theoretically this is given by the Richardson-Dushman equation:

$$j = 4\pi me(kT_{\rm d})^2 h^{-3} \exp[-\phi/kT_{\rm d}] = AT_{\rm d}^2 \exp[-\phi/kT_{\rm d}] \qquad (4.2)$$

for a solid at temperature $T_{\rm d}$; ϕ is the work function for the solid, and is typically a few electron volts for most solids of astronomical interest. As with the photoelectric effect the work function has to be modified if the grain is charged. Eq. (4.2) would lead us to expect that A is the same for all solids. However while Eq. (4.2) does describe the general character of thermionic emission quite well, there are in practice distinct differences in the measured values of A for different materials. This is largely due to the presence of impurities on the surface of the solid being investigated, which have a substantial effect on the thermionic current. It is customary to write Eq. (4.2) in the form

$$j = A(1-r)T_{\rm d}^2 \exp[-\phi/kT_{\rm d}], \qquad (4.3)$$

where $A = 4\pi mek^2/h^3 \simeq 1.2 \times 10^6$ ampere m^{-2}K^{-2} and r is a 'reflection coefficient' which takes account of electrons that have sufficient energy to clear the potential barrier at the surface but fail to do so. Typical values are $\phi \simeq 4.2$ eV and $\simeq 5$ eV for carbon and silicates respectively, and $A(1-r) \simeq 4 \times 10^4$ ampere m^{-2} K^{-2} for carbon.

4.2.3 Grain charge in a plasma

We now consider the charge acquired by a spherical grain immersed in a plasma which contains free electrons and ions; for simplicity, we assume that all the ions are identical. In such an environment the thermal energy will be equally divided between the electrons and ions so that, on the average, the mean electron kinetic energy is equal to the mean ion kinetic energy: $\frac{1}{2}m_{\rm e}v_{\rm e}^2 = \frac{1}{2}m_{\rm i}v_{\rm i}^2$, where subscripts e and i refer to electrons and ions respectively. The typical electron velocity $v_{\rm e}$ therefore exceeds the typical ion velocity by a factor $[m_{\rm i}/m_{\rm e}]^{1/2} \gg 1$. Unless the 'sticking probability' (defined below) for electrons is very much smaller than for ions more electrons than ions will stick and the grain will acquire a negative charge. However the negative charge on the grain will not increase indefinitely. As the grain charge increases the coulomb repulsion between grain and electrons will deflect all but the most energetic electrons; the ions on the other hand will be attracted. As a result the charge on the grain settles at an equilibrium value such that the charging and discharging effects of the electrons and ions are equal. Obviously the charge on the grain will be at all times an integral multiple Z of the electronic charge e.

We can make an order-of-magnitude estimate of the equilibrium charge Ze on a spherical grain of radius a by supposing that the electrons and ions all move with a speed corresponding to the average kinetic energy, as determined by the equipartition theorem:

$$\frac{1}{2}m_e v_e^2 = \frac{1}{2}m_i v_i^2 = \frac{3}{2}kT_{gas} \tag{4.4}$$

where T_{gas} is the temperature of the gas. At equilibrium, we might expect the impact rates of the electrons and ions to be equal. We must first determine the velocity with which the ion strikes the surface of a negatively charged grain. We use the principle of conservation of energy:

$$\frac{1}{2}m_i v_i^2 = \frac{1}{2}m_i V_i^2 - \frac{Ze^2}{4\pi\epsilon_0 a}. \tag{4.5}$$

The left hand side of Eq. (4.5) is the total energy of the ion at infinity, while the right hand side is the total energy at the moment of impact (V_i is the impact velocity). Conservation of energy thus gives the impact velocity as

$$V_i = \left\{\frac{2}{m_i}\left(\frac{1}{2}m_i v_i^2 + \frac{Ze^2}{4\pi\epsilon_0 a}\right)\right\}^{1/2}. \tag{4.6}$$

The *rate* at which ions strike the grain surface is $n_i\sigma V_i$, where n_i is the number of ions per unit volume and σ is the appropriate cross-section for collision. The impact rate

$$\mathcal{R}_i = n_i\sigma\left\{\frac{2}{m_i}\left(\frac{1}{2}m_i v_i^2 + \frac{Ze^2}{4\pi\epsilon_0 a}\right)\right\}^{1/2}.$$

The impact rate for electrons $\mathcal{R}_e$ is given by a similar expression, except that the '+' sign in the round brackets is replaced by a '−' sign and of course all 'ion' suffixes i are replaced by 'electron' suffixes e:

$$\mathcal{R}_e = n_e\sigma\left\{\frac{2}{m_e}\left(\frac{1}{2}m_e v_e^2 - \frac{Ze^2}{4\pi\epsilon_0 a}\right)\right\}^{1/2}.$$

At equilibrium we can equate the two expressions for $\mathcal{R}_i$ and $\mathcal{R}_e$ and, making use of Eq. (4.4), we can solve for Z to find

$$Z = \frac{3}{2}kT_{gas}\frac{4\pi\epsilon_0 a}{e^2}\left\{\frac{m_i - m_e}{m_i + m_e}\right\} \simeq \frac{3}{2}kT_{gas}\frac{4\pi\epsilon_0 a}{e^2} \tag{4.7}$$

if $n_i = n_e$; we have also made use of the fact that $m_i \gg m_e$. A more rigorous derivation, given in Appendix A, gives the result

$$Z = x_0 k T_{gas} \frac{4\pi\epsilon_0 a}{e^2}, \tag{4.8}$$

where x_0 is a constant of order unity, the value of which depends on the composition of the plasma; for example, $x_0 \simeq 2.5$ for a pure hydrogen plasma. In pure hydrogen at temperature 100 K (typical of the temperature of the interstellar gas), thermal velocities are low and in the absence of any other charging processes $Z \simeq 1$. On the other hand in the vicinity of a hot star, where the gas temperature may be 8000 K, we find $Z \simeq 120$.

It is also of interest to estimate the time taken for the grain to attain the equilibrium value of charge. We refer to Eqs (A.10) and (A.13), which give separately the rates at which negative and positive charge are accumulated by the grain. The difference between these two terms gives the total rate at which negative charge is accumulated, i.e. we have a differential equation for Ze:

$$\dot{Q} = \frac{d(Ze)}{dt} = |\dot{Q}_-| - |\dot{Q}_+|. \tag{4.9}$$

The time for the grain to reach equilibrium is approximately given by

$$t_{eq} \simeq \left|\frac{Q}{\dot{Q}}\right|.$$

Since the charging is dominated by electrons, by virtue of their smaller mass and higher velocity, we approximate $\dot{Q}$ by $\dot{Q}_-$, so that

$$t_{eq} \simeq x_0 \exp[x_0] \frac{4\pi\epsilon_0}{\pi a n_e e^2} \left(\frac{m_e k T_{gas}}{8\pi}\right)^{1/2} \tag{4.10}$$

where we have put $S_e = 1$, and $x_0 \simeq 2.5$ for pure hydrogen (see above).

If we take typical values for the interstellar medium, $T_{gas} \sim 10^2$ K and $a \sim 0.1\,\mu$m, we get $t_{eq} \simeq 3 \times 10^9/n_e$ sec. Thus even in the diffuse interstellar medium, where the number density of free electrons $n_e \sim 10^4\,\mathrm{m}^{-3}$, the grain reaches its equilibrium charge within a day or so, a timescale that is far shorter than that of most processes occurring in the interstellar medium. Around a hot star with $T_{gas} = 10^4$ K and $n_e = 10^7\,\mathrm{m}^{-3}$, $t_{eq} \simeq 10^3$ s. Thus in virtually all circumstances of interest the charging process is essentially instantaneous.

If we now add the photoelectric effect to the charging process we again need to balance the effects of accumulation of positive charge by ion collisions and the photoelectric effect, and the effects of accumulation of negative charge by electron collisions. However when we add the photoelectric effect to the discussion, we can take advantage of what we know about the charging of a grain in a plasma: the effect of electrons is far more important than that of ions in determining the grain charge. Accordingly we can proceed by ignoring the effect of ions and assume that positive charge is provided only by the photoelectric effect.

Although the impact of electrons and ions, and the photoelectric effect, are the most important processes determining grain charge in astrophysical environments, the thermionic effect may also be significant in certain circumstances for grains close to a hot star.

4.3 Temperature

As already noted, an important physical property of a cosmic dust grain is its temperature and, armed with the knowledge of how efficiently a grain absorbs radiation, we are in a position to calculate the temperature of a grain.

In astronomical environments there are several ways of heating a grain and each of these will be important or negligible depending on the circumstances. The principle of calculating the temperature of a grain is quite straightforward. We have a source of energy, which the grain absorbs at a known rate. The absorbed energy heats up the grain material so that it acquires a specific temperature. At this temperature the grain emits radiation and an equilibrium grain temperature is attained such that the power absorbed by the grain is precisely matched by the power emitted.

We have already seen in Chapter 3 how photons are converted to phonons when an electromagnetic wave interacts with a solid. The velocity of sound in a typical grain material is $\sim$ a few $\mathrm{km\,s^{-1}}$; cosmic dust grains generally have dimensions $\lesssim 1\,\mu\mathrm{m}$, so the absorbed energy is distributed throughout the grain in $\lesssim 10^{-8}$ s–essentially instantaneously for all practical purposes.

The most common sources of energy for grain heating in astrophysical situations are (i) radiation, (ii) impact of atoms or ions, (iii) chemical reactions on grain surfaces.

4.3.1 Grain heating by radiation

For a dust grain in interstellar space, or near a star, the most important mechanism of grain heating is by radiation. A grain placed near a source of radiation will absorb the radiation, the efficiency with which it does so depending both on the frequency of the radiation and on the nature of the grain, but generally speaking absorption will be most efficient at shorter (optical–UV) wavelengths for electrical conductors and close to frequencies like ω_{O2} for non-conductors (cf. the behaviour of Q_{abs} discussed in the previous chapter). In either case the grain gains thermal energy at the expense of the incident radiation: the absorbed radiation increases the temperature of the grain so that the grain (like any hot body) emits radiation, and an equilibrium temperature is attained at which the rate of absorption of energy is equal to the rate of emission. The principle can most easily be understood by considering an ideal, blackbody grain, which absorbs and emits with unit efficiency at all wavelengths. Take the case of a grain situated a distance r from a star of bolometric luminosity L_{bol} and temperature T_{eff}. We suppose that the star radiates like a blackbody, and may effectively be treated as a point source. If the star emits its radiation isotropically, then a spherical grain of radius a will intercept a fraction $\pi a^2/4\pi r^2$ of the star's radiation. Thus the power absorbed by the grain is simply

$$L_{bol}\frac{\pi a^2}{4\pi r^2}. \tag{4.11}$$

The grain attains an equilibrium temperature T_d, such that it emits as much radiation as it absorbs. At temperature T_d, a blackbody grain will emit radiation at a rate given by the Stefan-Boltzmann law of blackbody physics, namely σT_d^4 per unit area. Thus the total power emitted by the grain is given by

$$4\pi a^2 \sigma T_d^4. \tag{4.12}$$

Equating the power absorbed (4.11) to the power emitted (4.12) and rearranging gives the grain temperature

$$T_d = \left(\frac{L_{bol}}{16\pi r^2 \sigma}\right)^{1/4}. \tag{4.13}$$

Note that the grain temperature depends only on the luminosity of the star and on the grain's distance from the star in the blackbody case.

The situation for 'real' (as opposed to blackbody) grains is somewhat more involved, because the efficiency with which a grain absorbs and emits radiation is determined by the absorption efficiency Q_{abs}. Now we know that Q_{abs} depends strongly on frequency (see Figs. 3.9, 3.10) and so we have to determine the power absorbed at each frequency and then sum over all frequencies. The rate at which energy is input into the grain is in this case given by

$$\int_0^\infty \frac{L_\nu}{4\pi r^2}\pi a^2 Q_{abs}(a,\nu)d\nu \tag{4.14}$$

which is the equivalent of Eq. (4.11); recall that $\pi a^2 Q_{\rm abs}$ is the absorption cross-section of the grain. The power radiated by the grain is given by the equivalent of Eq. (4.12):

$$\int_0^\infty 4\pi a^2 Q_{\rm abs}(a,\nu) B_\nu(T_{\rm d}) d\nu. \tag{4.15}$$

Note that the grain behaves as though it has emissivity $Q_{\rm abs}(a,\nu)$. In this case the maximum emission occurs not at the wavelength $\lambda_{\rm max}$ given by Eq. (1.20) but at a wavelength given by[1]

$$\lambda_{\rm max} T = 2890 \left(\frac{5}{\alpha+5}\right). \tag{4.16}$$

Solving Eqs. (4.14) and (4.15) for $T_{\rm d}$ is generally rather cumbersome particularly if (as is often the case) the electrical conductivity depends on temperature; in general the determination of $T_{\rm d}$ has to be carried out numerically. However it is sometimes possible to extract $T_{\rm d}$ explicitly, as follows. First, let us define the *Planck mean* of the absorption efficiency $Q_{\rm abs}$. This is given by

$$\overline{Q_{\rm abs}} = \frac{\int_0^\infty Q_{\rm abs}(\nu,a) B_\nu(T) d\nu}{\int_0^\infty B_\nu(T) d\nu}. \tag{4.17}$$

This is simply the average of the absorption efficiency $Q_{\rm abs}$ over all frequencies, weighted according to the Planck function. [There are of course equivalent means for the other Q-factors.] Note that, since we are integrating over frequency, the dependence of $Q_{\rm abs}$ on frequency and grain radius becomes a dependence on *temperature* and grain radius for $\overline{Q_{\rm abs}}$. The dependence of $\overline{Q_{\rm abs}}$ on grain radius and temperature for spherical carbon grains is shown in Fig. 4.1.

Now multiply the top and bottom of Eq. (4.15) by $\int_0^\infty B_\nu(T_{\rm d}) d\nu = \sigma T_{\rm d}^4/\pi$. We then have

$$4\pi a^2 \overline{Q_{\rm abs}}(T_{\rm d}, a) \frac{\sigma T_{\rm d}^4}{\pi} \tag{4.18}$$

for the power radiated by the grain. Similarly we can rewrite Eq. (4.14) with the help of Eq. (1.8) and Eq. (1.19). If we suppose that a star radiates like a blackbody at temperature $T_{\rm eff}$ then we can replace $L_\nu/4\pi r^2$ in Eq. (4.14) by $B_\nu \pi R_*^2/r^2$. We now multiply Eq. (4.14) by $\int_0^\infty B_\nu(T_{\rm eff}) d\nu = \sigma T_{\rm eff}^4/\pi$. But since $L_{\rm bol} = 4\pi R_*^2 \sigma T_{\rm eff}^4$ the input of radiative energy into the grain is:

$$L_{\rm bol} \frac{\pi a^2 \overline{Q_{\rm abs}}(T_{\rm eff}, a)}{4\pi r^2}. \tag{4.19}$$

[1] As with Eq.(1.20), Eq. (4.16) refers to maximum flux per unit wavelength interval; for frequency units the numerical constant in Eq. (4.16) should be replaced by 5100 and the 5 in the round brackets replaced by a 3.

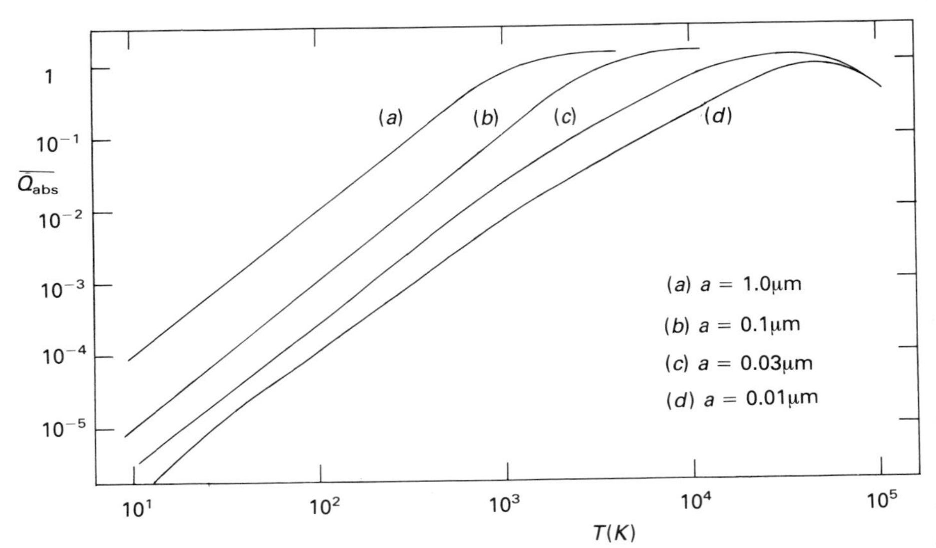

Figure 4.1: Planck mean absorption efficiency for spherical carbon grains.

As we have already seen, $Q_{\rm abs} \propto x^\alpha$ for small enough x, in other words $Q_{\rm abs} \propto \nu^\alpha$. For our present purposes we can take 'small enough x' to mean wavelengths in excess of a few μm–i.e. in the infrared. Now if $Q_{\rm abs}(\nu, a) \propto a\nu^\alpha$, the Planck mean $\overline{Q_{\rm abs}} \propto aT^\alpha$. Indeed, for a conducting grain material like carbon, the Planck mean of the absorption efficiency can often be expressed as the smaller of AaT^α and unity:

$$\overline{Q_{\rm abs}} = \min[1, AaT^\alpha], \tag{4.20}$$

where A is a constant (see Fig. 4.1). For example, with a in metres and T in degrees Kelvin, $A \simeq 1$ and $\alpha \simeq 2$ for graphitic carbon.

Equating Eqs (4.18) and (4.19) and solving for $T_{\rm d}$ we get

$$T_{\rm d} = \left(\frac{L_{\rm bol}}{16\pi r^2 \sigma} \frac{\overline{Q_{\rm abs}}(T_{\rm eff}, a)}{\overline{Q_{\rm abs}}(T_{\rm d}, a)} \right)^{1/4} . \tag{4.21}$$

We note that Eq. (4.21) gives the temperature of *grey* grains, i.e. grains whose absorption efficiency differs from unity but which is none-the-less independent of frequency (so that the Planck mean is independent of temperature). Note that the grain temperature again seems to be independent of grain radius, as it was in the blackbody case. However in Eq. (4.21) the grain radius is actually implicitly present (for the non-grey case) in the grain radius-dependence of $\overline{Q_{\rm abs}}$.

We now use Eq. (4.20) for the Planck mean and substitute in Eq. (4.21). Suppose we have a stellar temperature and grain size such that $Q_{\rm abs}(T_{\rm eff}, a) = 1$ while $Q_{\rm abs}(T_{\rm d}, a) = AaT_{\rm d}^{\alpha}$. A little rearrangement leads to

$$T_{\rm d} = \left(\frac{L_{\rm bol}}{16\pi r^2\sigma}\frac{1}{Aa}\right)^{1/(\alpha+4)}. \tag{4.22}$$

However for small grains, such that $AaT^{\alpha} < 1$ even at stellar temperatures, we have

$$T_{\rm d} = \left(\frac{L_{\rm bol}T_{\rm eff}^{\alpha}}{16\pi r^2\sigma}\right)^{1/(\alpha+4)}. \tag{4.23}$$

Note that, according to Eq. (4.22), a small grain is hotter than a larger grain at the same distance from the same source of radiation. However the temperature of *very* small grains is a topic to which we shall return in Section 4.3.4.

4.3.2 Grain heating by gas impact

In most cases, cosmic dust grains are heated radiatively but there are circumstances under which other sources of grain heating will be important. For example, interstellar dust grains are sometimes caught up in shocks arising as a result of supernova explosions. These shocks are typically at temperatures $\sim 10^7$ K and a dust grain in this sort of environment will be heated more effectively by the impact of the ions and electrons in the gas than by the interstellar radiation field. The other side of this particular coin is that the impact of ions also has the effect of eroding the grain away by the process of sputtering. In this section however we shall discuss the heating effects of a hot gas and defer a discussion of its destructive effects until later.

On a microscopic level, the heating effect of an impact by an atom, electron or ion is as follows; for brevity we shall refer in the following to ions only. The ion moves with a speed that is characteristic of the thermal energy of the gas, which has temperature $T_{\rm gas}$. For an ion of mass $m_{\rm i}$, the speed $v_{\rm i}$ will be given in terms of the gas temperature by Eq. (4.4)

$$\frac{1}{2}mv_{\rm i}^2 = \frac{3}{2}kT_{\rm gas}.$$

When the ion strikes the surface of the grain, it is likely to remain on the grain surface for a short time, which we can derive as follows. A grain surface is unlikely to correspond to that of a perfect crystal; it will have various defects, such as screw and edge dislocations, impurity atoms, vacancies etc. Each of these, as well as the component atoms of the grain surface, will present potential barriers which a surface ion must overcome before escaping the surface of the grain.

If an ion is trapped at a site where the potential well depth is V, and if it vibrates with frequency ν_0, then it will encounter the potential barrier $\sim \nu_0$ times per second. The probability that the ion will leave the site is proportional to the Boltzmann factor, $\exp[-V/kT_d]$, where T_d is the temperature of the grain. The frequency with which the ion leaves a site is given by $\nu_0 \exp[-V/kT_d]$. The *time* for which the ion remains at the site, τ_{well}, is therefore given by

$$\tau_{well} \simeq \nu_0^{-1} \exp\left[\frac{V}{kT_d}\right]. \tag{4.24}$$

The frequency of vibration of the ion on the grain surface ν_0 is typically $\sim 10^{13}\ s^{-1}$. The depth of a potential well at a site on the grain surface will obviously depend on the nature of the site. However in general we might expect $V \sim 0.02\,eV$. If we take a typical grain at temperature 20 K we find that $\tau_{well} \sim 10^{-6}$ sec. During this time the kinetic energy of the ion is imparted to the grain. The ion eventually leaves the surface of the grain with a speed which is considerably less than that with which it arrived; in fact its speed of departure v_f will be characteristic of the temperature of the dust grain rather than that of the gas. Thus

$$\frac{1}{2} m_i v_f^2 = \frac{3}{2} kT_d. \tag{4.25}$$

The ion has therefore provided a net amount of energy

$$E = \frac{1}{2} m(v_f^2 - v_i^2) = \frac{3}{2} k(T_{gas} - T_d) \tag{4.26}$$

to the grain. What we need however is the *rate* at which energy is supplied to the grain, so that we can match this up with the rate at which the grain then radiates away the absorbed energy. We therefore need to know the rate at which the ions strike the grain surface, and this depends not only on the ion speed but also on the number of ions per unit volume n_i in the gas.

In the simple case we could make the approximation that all the ions travel with speed v_i given by Eq. (4.4), in which case the number of ions striking the surface of a spherical grain of radius a per second is $4\pi a^2 n_i v_i$. If each ion gives up all its kinetic energy to the grain then the power delivered to the grain is just $4\pi a^2 n_i v_i \times \frac{1}{2} m v_i^2$. However the transfer of energy is unlikely to be completely efficient and if it occurs with efficiency ϵ (which must be < 1) then the rate at which energy is absorbed by the grain is $(4\pi a^2 \epsilon n_i v_i) \times (\frac{1}{2} m v_i^2)$.

As before, we equate the rate at which energy is absorbed by the grain to the rate at which the grain, at temperature T_d, emits energy in the form of radiation. If we suppose that the grain radiates like a blackbody, this is again $4\pi a^2 \sigma T_d^4$. This gives the grain temperature as

$$T_d \simeq \left[\left(\frac{3kT_{gas}}{m_i} \right)^{3/2} \frac{n_i m_i \epsilon}{2\sigma} \right]^{1/4} \tag{4.27}$$

provided $T_{gas} \gg T_d$, which is nearly always the case.

In deriving the result (4.27) we have made a number of approximations and assumptions, including (i) that the grain radiates like a blackbody, (ii) ignored the Maxwellian character of the velocity distribution of ions in the gas, (iii) neglected the fact that the grain is unlikely to be electrically neutral and (iv) that the ion actually 'feels' the surface of the grain. In a rigorous calculation, we should include not only the emissivity of the grain (in other words, its Q_{abs}) and the velocity distribution of the ions, but also the fact that in a hot gas, the grain is likely to be charged (see Section 4.2). For an incident ion this–as we have already seen–has the effect of modifying the cross-section of the grain: in the case of a negatively charged grain, a positive ion will see a grain cross-section that is larger than the geometrical cross-section because of coulomb attraction. Furthermore, for high enough gas temperatures, electrons are quite likely to pass straight through the grain without even knowing it is there, so that the effect of electron heating is correspondingly reduced. Nonetheless the result (4.27) will give a reasonable indication of the efficacy of a hot gas in heating a dust grain.

We can take some of the above factors into account in a more rigorous derivation. Suppose a positive ion of velocity v_i strikes the surface of a negatively charged grain. If all the kinetic energy available finds its way into the grain the energy input in the course of the collision is just $\frac{1}{2}m_i v_i^2$; in reality of course only a fraction ϵ of the kinetic energy is available. The fraction of atoms or ions having velocity in the range $v_i \rightarrow v_i + dv_i$ is given by the Maxwell distribution (E.1). Thus the total energy going into the grain per unit time is

$$E_{in} = [\sigma_i S_i f(v_i) n_i v_i] \times [\frac{1}{2} m_i v_i^2 \epsilon] dv_i, \tag{4.28}$$

where σ_i is given by Eq. (A.3), n_i is the number density of ions and S_i is the sticking probability. As noted above the ion leaves the grain surface at a velocity that is characteristic of the grain (rather than gas) temperature so the power lost as ions leave is

$$E_{out} = [\sigma_i S_d f(v_d) n_i v_d] \times [\frac{1}{2} m_i v_d^2] dv_d, \tag{4.29}$$

where the subscript d emphasizes the fact that the atom or ion is coming off at a velocity consistent with the temperature of the dust grain. At the moment we are interested only in the heating effects of the gas and not in the growth and destruction of the grain as a result of being immersed in the gas. To get the total power input, by ions of all velocities, we suppose that the *net* flux of incoming and outgoing ions is zero, i.e. $\sigma S_i n_i v_i - \sigma S_d n_d v_d = 0$. We denote by $\mathcal{R}$ the quantity $\sigma S_i n_i v_i$. The *net* power input by ions with velocities in the range $v_i \to v_i + dv_i$, E'_{in}, is simply the difference between E_{in} and E_{out}:

$$E'_{in} = \mathcal{R}\left[\frac{1}{2}m_i v_i^2 f(v_i)dv_i - \frac{1}{2}m_d v_d^2 f(v_d)dv_d\right]. \tag{4.30}$$

To get the *total* power input we integrate each of the terms in Eq. (4.30) over the Maxwell velocity distribution (E.1). The first term gives $\frac{3}{2}\mathcal{R}kT_{gas}$, while the second term is $\frac{3}{2}\mathcal{R}kT_d$. The quantity $\mathcal{R}$ is determined for a neutral grain in Chapter 5 [see Eq. (5.10)]; in the present case, however, we need to take into account the increased cross-section of a negatively charged grain for collision with a positive ion. We therefore have

$$\mathcal{R} = n_i \int_0^\infty \sigma_i v_i f(v_i) dv_i.$$

Substituting for σ_i from Eq. (A.3) and for the Maxwell distribution (E.1) we get

$$\mathcal{R} = \pi a^2 n_i \left(\frac{2kT_{gas}}{\pi m_i}\right)^{1/2}\left(1 + \frac{Ze^2}{4\pi\epsilon_0 akT_{gas}}\right);$$

the integration over v_i is easily solved by the substitution $x = m_i v_i^2/2kT_{gas}$. The net input of power by ion collision is therefore given by

$$E_{in} = \frac{3\pi a^2}{2} S\epsilon n_i \left[\frac{2kT_{gas}}{\pi m_i}\right]^{1/2} kT_{gas}\left[1 + \frac{Ze^2}{4\pi\epsilon_0 akT_{gas}}\right](1 - T_d/T_{gas}). \tag{4.31}$$

In a hot gas we would expect both ions and electrons to contribute to grain heating. The effect of electrons is obtained by simply following through the derivation that led to Eq. (4.31), and noting that the cross-section for the collision of electrons with a negatively charged grain is given by Eq. (A.4). The combined effect is then obtained by simply summing the ion and electron values of E_{in}.

Notice from Eq. (4.31) how we need $T_{gas} > T_d$ to get a net input of energy from the gas into the grain, as we would expect intuitively. While this is in fact the case the gas has, in general, to be extremely hot before the heating effects of the gas begin to compete with radiative heating. To see this let us look at a blackbody grain, firstly at distance r from a star of bolometric luminosity $L_{\rm bol}$, and secondly, in interstellar space, with no nearby star to dominate the radiative heating. In the first case the radiative power input into a grain of radius a is given by Eq. (4.11), whereas in the second case the power input is $4\pi a^2 \Psi \sigma T_{\rm is}^4$, where $\Psi(\sim 10^{-14})$ and $T_{\rm is}$ are respectively the 'dilution factor' and effective temperature for the heating radiation (see Section 6.6.1 below).

In the case of circumstellar dust, we have

$$\frac{E_{\rm in}}{E_{\rm cs}} \simeq 1.3 \times 10^{-22} \frac{n_{\rm i}}{L/L_\odot} T_{\rm gas}^{3/2} \left(\frac{r}{10^{12}{\rm m}} \right)^2$$

where, as before, r is the star-grain distance and we have set $S = 1$. $E_{\rm cs}$ is the radiative power input, given by Eq. (4.11). For heating by gas to be significant, we need $E_{\rm in}/E_{\rm cs} > 1$; thus for $L = L_\odot$ and $r = 10^{12}$ m, $n_{\rm i} T_{\rm gas}^{3/2}$ must be $\gtrsim 7.6 \times 10^{21}$ m^{-3}-K$^{3/2}$. Obviously impact heating will dominate or not depending on the environment but if, for example, $n_{\rm i} \sim 10^{12}$ m^{-3} then $T_{\rm gas}$ must exceed 3.9×10^6 K. Similarly in interstellar space we need

$$\frac{E_{\rm in}}{E_{\rm is}} \simeq 1.8 \times 10^{-16} n_{\rm i} T_{\rm gas}^{3/2},$$

where $E_{\rm is}$ is again the radiative power input. In interstellar space the gas density $n_{\rm i} \sim 10^5$ m^{-3} so the gas temperature needs to be $\gtrsim 15 \times 10^7$ K before heating by impact begins to dominate.

In either case then the gas must be at a temperature of several millions of degrees to be a major contributor to dust heating. Gas at these temperatures is found in the remnants of supernova explosions, following what is essentially the instantaneous input of a large amount of energy ($\sim 10^{43}$ J) into the surrounding medium. This event results in a shock which heats the gas to temperatures $\sim 10^6$ K, at which temperature the gas emits at X-ray wavelengths. Any interstellar dust grains which are swept up in the shock will therefore be heated by the impact of ions in the hot gas to temperatures ~ 30 K; such emission was detected by the *IRAS* satellite. Fig. 4.2 shows the X-ray emission from a supernova remnant in the Large Magellanic Cloud, together with the position of the corresponding *IRAS* source. In this case the infrared luminosity is about 12× the X-ray luminosity, and the mass of the infrared emitting dust is about 0.6 % of the X-ray emitting gas, as would be expected for a gas-to-dust ratio similar to that appropriate for the Galaxy.

Note that the gas is depositing thermal energy in the dust, which is then heated up and radiates away the absorbed energy in the infrared. At

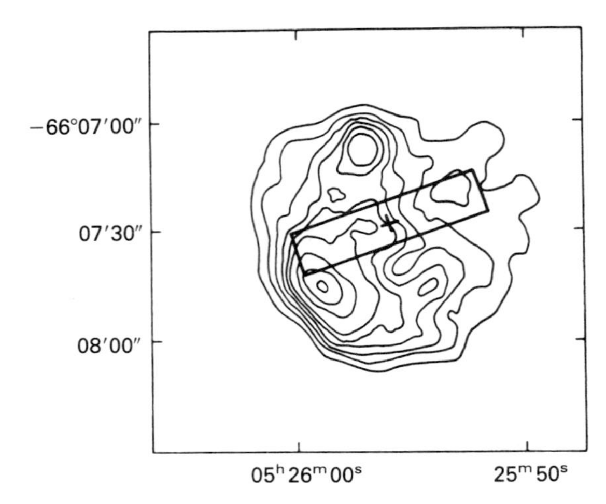

Figure 4.2: X-ray emission from the supernova remnant N49 in the Large Magellanic Cloud (contour lines) and the position of the corresponding infrared source detected by *IRAS* (rectangle). After J. R. Graham et al., *Astrophysical Journal*, Vol. **319**, 126 (1987).

higher temperatures ($\gtrsim 10^7$ K) electrons begin to penetrate the grains and their effect in grain heating is consequently diminished. Nevertheless this mechanism is very efficient at cooling a hot gas and indeed, at temperatures above about 3×10^6 K the cooling of a gas by this process is more efficient than cooling by (for example) the emission of spectral lines (see Fig. 4.3).

4.3.3 Grain heating by chemical reaction

The third mechanism of grain heating is by the effects of chemical reactions on the grain surface; a necessary condition for this to happen of course is that the reaction be *exothermic*, i.e. the reaction must supply, rather than require, energy to proceed.

We have already discussed the discharging of a negative grain by the impact of ions. In the case of hydrogen we have

$$H^+ + e^- \overset{grain}{\rightarrow} H + E,$$

where E is the energy available from the reaction. Another important grain surface reaction is the formation of hydrogen molecules H_2. Indeed, since gas densities in interstellar space are so low the only plausible route for hydrogen molecule formation in interstellar space is on the surfaces of dust grains, via

$$H + H \overset{grain}{\rightarrow} H_2 + E. \tag{4.32}$$

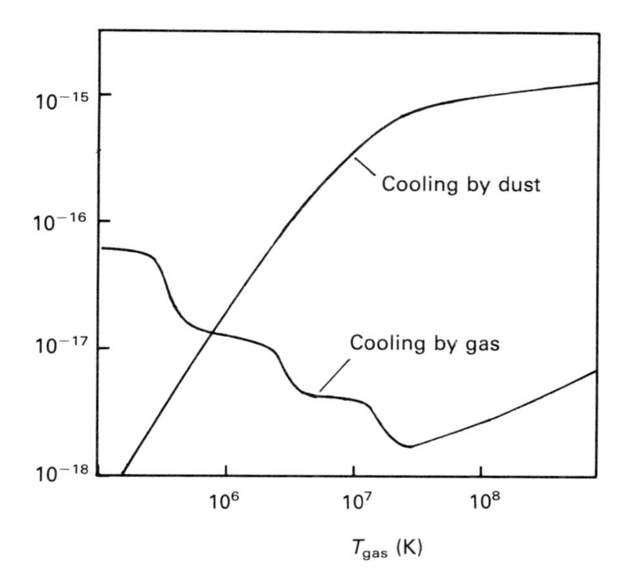

Figure 4.3: Efficiency of cooling a hot gas by line emission and by infrared emission by dust grains heated by the hot gas. Ordinate represents the power emitted per unit volume, normalized to unit number density; scale arbitrary. Adapted from B. T. Draine, *Astrophysical Journal*, Vol. **245**, 880 (1981).

In the former case, $E \leq +13.6$ eV, while in the latter case $E \leq +4.48$ eV. In both cases the reaction is exothermic. As the reaction is occurring on the surface of a grain some of this energy may be available to penetrate into the body of the grain to heat it up.

Let us look at the formation of molecular hydrogen on a grain surface. A necessary condition for the reaction (4.32) to proceed is that there are at least two hydrogen atoms on the grain surface. When a hydrogen atom in the gas phase lands on the grain surface, there is a probability $S_{\rm H}$ that it will 'stick' rather than immediately bounce back into the gas. Once on the surface a hydrogen atom will remain there for a time determined by Eq. (4.24), with V again $\simeq 0.02\,\text{eV}$.

While the atom is on the grain surface a second hydrogen atom must arrive, so that the reaction (4.32) can go ahead. On a spherical grain of radius a, immersed in a gas at temperature $T_{\rm gas}$, the rate at which hydrogen atoms strike the surface is given by Eq. (5.11) below. We might expect therefore that H_2 molecules will form provided that

$$T_{\rm d} < \frac{V}{k}\left\{\log_{\rm e}\left(\frac{\nu_0}{4\pi a^2 n_{\rm H}}\left[\frac{kT_{\rm gas}}{2\pi m_{\rm H}}\right]^{-1/2}\right)\right\}^{-1}. \tag{4.33}$$

Suppose we take $a \simeq 0.1\ \mu$m, $T_{\rm gas} \simeq 100$ K and $n_{\rm H} \simeq 10^5$ m^{-3} as being fairly typical values. Then with $\nu_0 \simeq 10^{13}$ Hz and $V/k \simeq 250$ K, (corresponding to $V \sim 0.02$ eV) we find that $T_{\rm d}$ has to be less than about 20 K for the formation of hydrogen molecules. This seems rather a stringent requirement, and perhaps hydrogen atoms are more tightly bound to the grain surface than our estimate of V would imply.

Even so, only a fraction γ_1 of atoms will react and in each reaction, only a fraction γ_2 of the available energy will go into the grain; the remainder will go into the departing hydrogen molecule, which is likely to be in excited vibrational and/or rotational states. For a spherical grain of radius a, the rate at which energy is input into the grain is

$$4\pi a^2 \gamma_1 \gamma_2 E S_{\rm H} n_{\rm H} \left(\frac{2kT_{\rm gas}}{m_{\rm H}}\right)^{1/2} . \tag{4.34}$$

We again equate this to the rate at which the grain radiates away the energy, according to Eq. (4.15) and consider the simple case of a blackbody grain for simplicity; also, to get an idea of the maximum effect, we set $\gamma_1 = \gamma_2 = S_{\rm H} = 1$. As we have seen above, hydrogen molecule formation is likely to occur on grains having $T_{\rm d} \lesssim 20$ K. Such grains are likely to be found in molecular clouds, the interiors of which are opaque to short-wavelength ($\sim$ ultraviolet, optical, and even near infrared) radiation so that radiative heating is unimportant. The densities in these clouds are $n_{\rm H} \sim 10^{10}$ m^{-3} (although this is already mostly in molecular form). With $E = 4.48$ eV in this case we find, on equating (4.34) to $4\pi a^2 \sigma T_{\rm d}^4$ that the resulting $T_{\rm d} \simeq 3.5$ K. Apparently then molecule formation can be a significant source of grain heating in opaque clouds.

In a dusty HII region, the gas temperature $T_{\rm gas} \simeq 10^4$ K, while the gas density $n_{\rm i} \simeq 10^7$ m^{-3}. Hydrogen recombination on the grain surface gives $E = 13.6$ eV and, proceeding as before, we find $T_{\rm d} \simeq 1.5$ K. Clearly the effect of grain heating in a hot gas by electron-proton recombination is negligible by comparison with radiative heating.

4.3.4 Do grains have an equilibrium temperature?

We pause for a moment here to ask whether grains actually do have an equilibrium temperature, as given for example by Eq. (4.21). Consider what implicit assumptions have been made in the derivation of Eq. (4.21). It is assumed that there is a continuous flux of radiation and that the energy supplied by an individual photon is negligible by comparison with the thermal energy content of the grain. Before looking at the conditions under which this assumption is justified, let us look at the consequences if they are *not* satisfied.

Suppose we have a grain of radius a at temperature T_d, and that it is shielded from radiation or any other source of heating. The thermal energy content of the grain is given simply by $U = C_V T_d$ per mole, where C_V is the specific heat (at constant volume) of the solid. At sufficiently high temperatures C_V is given by the classical value $3\Re$ J mole^{-1} deg^{-1} (the Dulong-Petit law), where $\Re$ is the universal gas constant. Since the grain has a definite temperature it will radiate–i.e. its thermal energy content will decrease, as will its temperature. If the grain behaved like a perfect blackbody it would emit radiation at a rate given by the Stefan-Boltzmann law, i.e. $\dot{U} = -4\pi a^2 \sigma T_d^4$ and the temperature would fall exponentially with a characteristic timescale

$$\tau_{\rm cool} \simeq \left|\frac{U}{\dot{U}}\right| = \frac{N_A k}{\sigma}\frac{a\rho}{\mu T_d^3}, \tag{4.35}$$

where N_A is Avogadro's number and μ is the molecular weight of the grain material of bulk density ρ. For $\mu \simeq 20$, $a \simeq 1\mu$m and $T_d \simeq 100$ K, fairly typical for a cosmic dust grain, $\tau_{\rm cool}$ is of the order of a few seconds.

We now allow an input of energy E to the grain, by a single isolated event. This event will most likely be the absorption of a single photon (of energy $E = h\nu$), but may also be an impact by a single gas atom or ion (Section 4.3.2) or the effect of a single chemical reaction (Section 4.3.3). In order to judge the effect of an isolated heating event we must compare E with the thermal energy content of the grain, as determined above. If E is very much greater than U the absorption of a single photon will have a substantial effect on the grain temperature, i.e. if

$$\left(\frac{a}{1\mu m}\right)^3 T_d \lesssim 1.5 \times 10^{-9}\left(\frac{E}{1\text{eV}}\right)\frac{\mu}{\rho}. \tag{4.36}$$

We see that the effect of a single heating event will be significant for small, cool grains. For example, if $a \simeq 0.001\,\mu$m, $\mu \simeq 20$ and $\rho \simeq 3000\,\text{kg}\,\text{m}^{-3}$ then $T_d \lesssim 100$ K for a 10 eV photon. At these temperatures the Dulong-Petit law is inappropriate; we should take into account the temperature-dependence of the specific heat as given, for example, by the Debye theory. In this case $C_V \propto T_d^3$ at low temperatures and only approaches the Dulong-Petit value at temperatures well in excess of the Debye temperature Θ for the solid. The effect of this is to make the limit (4.36) even more severe.

A further factor is the relative magnitudes of the cooling time and the time interval $\tau_{\rm int}$ between individual heating events. If $\tau_{\rm int} > \tau_{\rm cool}$ then the grain has ample time to cool significantly between heating events, and the

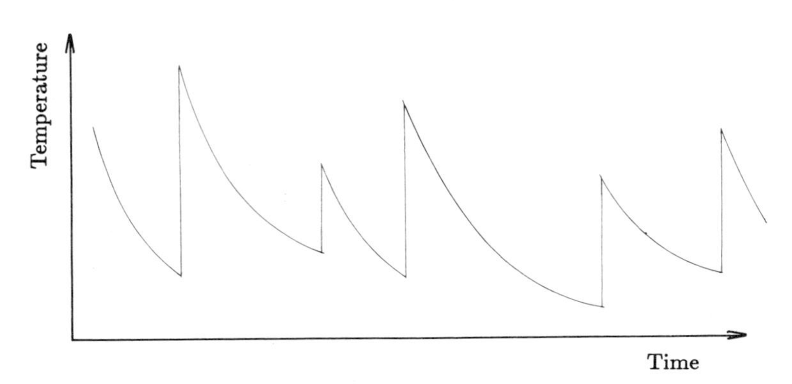

Figure 4.4: Variation of the temperature of a dust grain subject to stochastic heating.

grain can not be regarded as having an equilibrium temperature; the grain temperature will consequently vary erratically as depicted in Fig. 4.4. The heating of the grain in this situation is essentially a statistical process and is sometimes referred to as *stochastic heating.*

Note that the peak temperatures correspond to the energy involved in each individual heating event, and that each of these peaks is at a substantially higher temperature than the equilibrium value which would have been obtained using the methods outlined in the previous sections. This process is likely to be of significance for small grains in interstellar space, where passage of a short-wavelength ($\sim$ optical-ultraviolet) photon is a pretty rare event.

4.4 Viscous drag

There are a number of circumstances in which a grain will be in motion relative to the surrounding gas. For example when a grain forms in a stellar wind (see Section 7.5.3) it is quite possible that, as a result of radiation pressure, it acquires a velocity different from that of the gas from which it formed. Under these conditions the motion of the grain will be subject to viscous forces, which we now briefly discuss.

Suppose we have a spherical grain, radius a, moving with velocity v through a gas at temperature $T_{\mathbf{gas}}$ and in which the number density of atoms is n. The viscous drag force experienced by the grain depends on whether the grain velocity is sub- or super-sonic, i.e. on whether

$$v < \left(\frac{kT_{\rm gas}}{m_{\rm X}}\right)^{1/2} \qquad \text{(sub - sonic)}$$

or

$$v > \left(\frac{kT_{\rm gas}}{m_{\rm X}}\right)^{1/2} \qquad \text{(super - sonic)}.$$

In the former case the drag force $F_{\rm drag}$ is given, for a spherical particle, by *Stokes' law*

$$F_{\rm drag} = 6\pi\eta a v, \tag{4.37}$$

where η is the viscosity. In terms of the gas properties, this is

$$\eta = \frac{1}{2} a n_{\rm X} m_{\rm X} \left(\frac{kT_{\rm gas}}{\pi m_{\rm X}}\right)^{1/2}, \tag{4.38}$$

where $m_{\rm X}$ is atomic (or molecular) mass of the gas. If the grain velocity relative to the gas is super-sonic then

$$F_{\rm drag} \simeq \pi a^2 n_{\rm X} m_{\rm X} v^2. \tag{4.39}$$

Note that, in the sub-sonic case, $F_{\rm drag} \propto v$, whereas $F_{\rm drag} \propto v^2$ in the super-sonic case.

4.5 Poynting-Robertson effect

A dust grain close to a source of radiation (in the case of solar system dust, the source is of course the Sun) will be illuminated anisotropically (i.e. from the direction of the source) but will re-emit the absorbed radiation isotropically. Because of this 'imbalance' the grain loses angular momentum and slowly spirals in towards the source.

Suppose we have a grain of radius a, moving around a source of radiation in a circular orbit of radius R with velocity V; the absolute value of the grain's orbital angular momentum is consequently $\mathcal{L} = m_{\rm g} V R$. A photon of frequency ν (and hence of momentum $h\nu/c$) is emitted by the source and is absorbed by the grain. Owing to aberration an imaginary observer on the grain sees the radiation coming from a direction that makes an angle $\theta = \cos^{-1}(-V/c)$ with the normal to the direction of motion; note that this is in the *opposite* direction to the grain's orbital motion. The angular momentum of the grain therefore decreases by an amount

$$\Delta\mathcal{L} = \frac{h\nu}{c} R\cos\theta = -\frac{h\nu}{c^2} RV \tag{4.40}$$

and the grain spirals in towards the source of radiation. To determine the combined effect of all the photons emitted by the source of radiation we need to sum over all frequencies. The rate at which the source emits photons of frequency ν is $L_\nu/h\nu$, where L_ν is the luminosity of the source at frequency ν and of these a grain of absorption cross-section $\sigma_{\rm g}$ intercepts a fraction $\sigma_{\rm g}/4\pi R^2$. Each of these photons reduces the grain's orbital angular momentum by an amount given by Eq. (4.40); consequently the total rate at which orbital angular momentum is lost is

$$\dot{\mathcal{L}} = -\frac{\sigma_{\rm g}}{4\pi R^2}\int_0^\infty \frac{L_\nu}{h\nu}\frac{h\nu}{c^2}RV\,d\nu. \tag{4.41}$$

Thus

$$\dot{\mathcal{L}} = -\frac{L_{\rm bol}}{4\pi R^2}\frac{\sigma_{\rm g}}{m_{\rm g}c^2}\mathcal{L}, \tag{4.42}$$

where $L_{\rm bol}$ is the bolometric luminosity of the source.

Although we have considered the absorption of radiation by a grain moving in a circular orbit, the result (4.42) also applies to isotropic scattering, and also to a grain moving in a non-circular orbit provided we take only that component of orbital velocity that is orthogonal to the source-grain vector (thus a grain falling directly towards the source of radiation, or moving directly away, is not subject to Poynting-Robertson drag). However note that, if the grain velocity has a component along the line joining the particle and the source of radiation, account should be taken of the fact that, from the point of view of an imaginary observer on the grain, the radiation from the source will be Doppler shifted.

From Eq. (4.42) we can see that the time taken for a particle to spiral in towards the source of radiation is roughly

$$\tau_{\rm PR} \sim |\mathcal{L}/\dot{\mathcal{L}}| = \frac{4\pi R^2}{L_{\rm bol}}\frac{m_{\rm g}c^2}{\sigma_{\rm g}}.$$

We can therefore estimate the time taken for a typical interplanetary grain (of radius $a = 1\ \mu$m and density $2000\,{\rm kg\,m^{-3}}$), moving in a circular orbit in the asteroid belt ($R \simeq 2.8$AU) to fall in towards the Sun. With $L_{\rm bol} = L_\odot$ we have $\tau_{\rm PR} \simeq 4.4 \times 10^4$ yrs. This is obviously negligible by comparison with the age of the solar system ($\sim 4.6 \times 10^9$ yrs).

Poynting-Robertson drag is a consequence of the interaction of radiation with a particle. In practice of course there are competing processes–some of which also arise from the interaction of the grain with radiation (see Section 3.4)–which affect the motion of the grain. In particular radiation will exert an outward force on a grain (due to radiation pressure) and this will tend to oppose the effect of Poynting-Robertson drag. Note however that the force on the grain arising as a result of radiation pressure will be directed radially away from the source of radiation, whereas that arising as a result of Poynting-Robertson drag will be in the opposite sense to the particle's instantaneous velocity vector. The resultant effect of these two forces will of course be the vector sum of the individual components.

Problems

4.1. Use Eq. (4.13) to determine the temperature of the Earth, assuming that the Earth-Sun distance is 1 AU (150×10^6 km) and that both Earth and Sun behave like blackbodies.

4.2. In deriving Eq. (4.13) it was assumed that the star is a point source. In some cases, the grain is close enough to the star that this assumption does not hold (i.e. the star shows a sizable disc as seen from the grain). How would this affect the result (4.13)?

4.3. When we derived the efficiency factor for radiation pressure $Q_{\rm pr}$, the comment was made that absorbed radiation is lost to the forward beam and plays no further part in the discussion. Yet the absorbed energy is re-radiated, so why does the re-emitted radiation (like the scattered radiation) not contribute to the forward momentum? Under what circumstances might the effect of re-emitted radiation be important in this context?

4.4. Deduce the conditions that a grain has a meaningful equilibrium temperature in the case of (i) heating by gas impact and (ii) heating by chemical reaction.

4.5. Using the radiation density in interstellar space (see Section 6.6.1), estimate the rate at which photons strike the surface area of a spherical interstellar grain, of radius 0.15 μm.

4.6. A purely absorbing grain of radius $a = 0.5\,\mu$m and density $3000\,\rm kg\,m^{-3}$ is situated a distance 1 AU from a star of luminosity $L_\odot$ and mass $M_\odot$. Determine the relative strengths of gravity and radiation pressure on the grain.

Reading

The book by Dyson & Williams is an undergraduate text that deals with molecular and dynamical processes in the interstellar medium, and contains complementary material for the present chapter:

[C] *Physics of the Interstellar Medium*, J. E. Dyson & D. A. Williams, Manchester University Press (1980).

The following paper provide a very full discussion of the physics of dust grains in a hot gas:

[D] B. T. Draine & E. E. Salpeter, *Astrophysical Journal*, Vol. **231**, 77 (1979).

5

Grain formation and destruction

5.1 Introduction

We shall see in Chapter 6 that interstellar grains can *not* form in interstellar space: grains must form in stellar atmospheres and then be ejected into the interstellar medium. Indeed grain formation can be seen to occur in the vicinity of certain stars in real time. This is not to say that, once in the interstellar medium, grains do not evolve and change. On the contrary the nature of a grain can change significantly once it finds itself in the interstellar medium. This change in the character of a grain can result from additional growth (such as the acquisition of a mantle of ice in a molecular cloud) or destruction, for example in the aftermath of a supernova explosion. In this chapter we look at some of the principles involved in understanding the processes of grain formation and destruction.

5.2 Change of phase

Grain nucleation, growth and destruction all involve a change of phase, from the gas phase to the solid phase or *vice versa*. Such changes of phase are usefully described by means of a phase diagram, such as that shown in Fig. 5.1. The various regions of Fig. 5.1 show the conditions under which the gas, liquid and solid phases can exist. Note that there is a point, the triple point, where all three phases can coexist; also there is a point, the critical point, above which the liquid phase can not exist. We should note that Fig. 5.1 shows just about the simplest phase diagram possible. Many materials can exist in different solid forms (for example amorphous carbon,

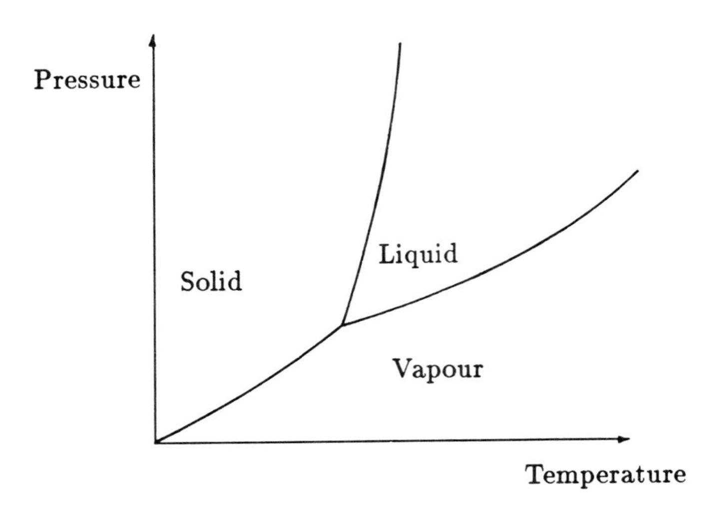

Figure 5.1: Simple phase diagram.

diamond, graphite, carbyne in the case of carbon) and each of the solid phases occupies different regions of the phase diagram. However this complication need not concern us here.

Suppose we have an ideal gas, at pressure P_{gas} and volume V. If we compress the gas at constant temperature (i.e. isothermally) its volume will initially change in such a way that PV = constant. However at a certain pressure, the gas will condense, whether into a liquid or solid depending on whether the temperature is greater than, or less than, the temperature corresponding to the triple point. However there is a specific condition which must be satisfied for condensation to occur. Suppose we have a chemical species X (where X might be C, or SiC etc.) in solid form at temperature T_{d}, situated in a vacuum. Atoms or molecules[1] of X will leave the surface and enter the vapour phase and eventually equilibrium is established between the solid and the vapour. The pressure of this vapour, for an infinite plane solid, depends on (amongst other factors) temperature. Thermodynamic arguments suggest that, for equilibrium between two phases, the specific Gibbs free energy (i.e. the Gibbs free energy per unit mass) must be the same for the two phases and, assuming that mass is conserved during the phase change, the *total* Gibbs function must be a minimum. This condition

[1] Henceforth we shall refer only to atoms, to avoid repetition of the cumbersome phrase 'atoms or molecules'.

Table 5.1: Parameters for astronomically relevant materials

Material	P_0 (N m^{-2})	T_0 (K)	γ (N m^{-2})	ρ (kg m^{-3})
Graphite	1.68×10^{13}	88880	$1 \rightarrow 3$	2200
Silicates	5.31×10^{13}	60560	0.9	3200–3500
Water ice	2.16×10^{5}	6160	...	1000
Hydrogen	2.66×10^{7}	104	...	70

leads to the Clausius-Clapeyron equation[2], which we write in the form

$$P_{\text{vap}} \simeq P_0 \exp[-T_0/T], \tag{5.1}$$

where P_0 and T_0 $(= \mathcal{L}/\Re)$ are constants the values of which are specific to each material; $\mathcal{L}$ is the latent heat of evaporation per mole. Some typical values of P_0 and T_0 are given in Table 5.1. Values for hydrogen are included in Table 5.1 for use in Chapter 9.

Now whether or not condensation from the gas to the solid phase will occur depends on the relative values of the gas pressure P_{gas} and the vapour pressure P_{vap} given by Eq. (5.1). Once P_{gas} exceeds P_{vap} then in principle, condensation should occur. In practice, of course, the relevant quantity is the *partial* pressure of the condensing species, not the *total* gas pressure; furthermore it is found that P_{gas} has to exceed P_{vap}, sometimes by a substantial amount, before condensation can occur: the gas must be *supersaturated.*

We should pause here for a moment and consider the relevance of Eq. (5.1): should we really be using the vapour pressure expression appropriate for an infinite plane surface? In fact there is a correction term to Eq. (5.1) which allows for a spherical surface of radius a:

$$P_{\text{vap}}(T, a) = P_{\text{vap}}(T, \infty) \exp\left[\frac{2\gamma m_{\text{X}}}{\rho k T a}\right], \tag{5.2}$$

where γ is the surface energy, m_{X} is the molecular mass of species X and ρ is the bulk density of the solid; values of γ and ρ are also given in Table 5.1. In Eq. (5.2) $P_{\text{vap}}(T, \infty)$ is just the vapour pressure given by Eq. (5.1). If we take values of γ and ρ from Table 5.1 and take carbon as a typical example we can easily estimate the effect of the curvature of the grain surface. The importance of the correction term in Eq. (5.2) is determined by the size of the

[2]The Clausius-Clapeyron equation is a limiting case of a more general equation, the Sackur-Tetrode equation, which is derived on the basis of more rigorous thermodynamic arguments than the Clausius-Clapeyron equation.

argument of the exponential term: if this begins to get close to, or greater than, unity the correction is important. Inserting values from Table 5.1 we see that the effect of curvature for a grain of temperature T_d will be important if its radius

$$a \lesssim \frac{1.3}{T_d} \ \mu m.$$

Evidently grains must be extremely small before the effects of surface curvature become important.

It is worth noting how (again) we must rely on laboratory work to address an astronomical problem: in order to investigate the conditions under which grains might condense in astronomical environments, we first need laboratory determinations of the phase diagrams, surface energies and densities for relevant materials.

5.3 Grain nucleation

In the previous section we spoke of the gas condensing into the liquid or solid phase, as though this were a phenomenon that occurs automatically. However before condensation can commence there must exist a 'seed'–a *nucleation site*–on which condensation can occur. This is a common requirement in a variety of situations, e.g. the growth of crystals in the laboratory, the formation of raindrops and of bubbles in a liquid and, from our point of view, the formation of a grain from the gas phase. In the astronomical situation, once the nucleation site is available there is no difficulty in understanding grain growth: it is the process of nucleation that is difficult to understand. Nucleation may occur via a variety of mechanisms but in general they come under two broad headings, *homogeneous* nucleation and *heterogeneous* nucleation.

Suppose again that we have a material X (such as carbon), which exists in the gas phase. Homogeneous nucleation consists of a series of reactions

$$\begin{aligned} \mathrm{X} + \mathrm{X} &\longleftrightarrow \mathrm{X}_2 + E_1 \\ \mathrm{X} + \mathrm{X}_2 &\longleftrightarrow \mathrm{X}_3 + E_2 \\ &\vdots \\ &\vdots \\ \mathrm{X} + X_\mathrm{n} &\longleftrightarrow \mathrm{X}_{\mathrm{n}+1} + E_\mathrm{n} \end{aligned}$$

and so on. In each reaction an amount of energy E_i, corresponding largely to the kinetic energy of the reacting species X and X_i, has to be dissipated; sometimes this can be done by the emission of one or more photons (the product being produced in an excited state). Note that only species X is involved in the process, hence the terminology *homogeneous* nucleation. In the first reaction however this is not, in general, possible: the product X_2 is homonuclear and shedding the excess energy radiatively is not allowed by the selection rules of quantum theory. In order that the first reaction can proceed we generally have to rely on three-body reactions

$$\mathrm{X} + \mathrm{X} + \mathrm{X} \longleftrightarrow \mathrm{X}_2 + \mathrm{X} + E_1',$$

the function of the third X being to carry off the energy of the reaction.

As the various reactions in the above sequence proceed, in both directions, we will eventually have an equilibrium situation in which the abundance n_i of the species $\mathrm{X_i}$ is given by

$$n_\mathrm{i} = n_1 \exp[-\Delta G_\mathrm{i}/kT]. \tag{5.3}$$

ΔG_i is the Gibbs free energy of formation of a cluster of radius a consisting of i atoms of X and may be determined by constructing the Gibbs function for the solid phase, the vapour phase and the surface, and requiring that the total Gibbs function is a minimum (cf. above). The result is

$$\Delta G_\mathrm{i} = 4\pi a^2 \gamma - \frac{4\pi a^3 \rho}{3} \frac{kT}{m_\mathrm{X}} \log_\mathrm{e} \mathcal{S} \tag{5.4}$$

for a spherical cluster, where $\mathcal{S}$ is the supersaturation ratio $P_\mathrm{gas}/P_\mathrm{vap}$. [Eq. (5.2) is a byproduct of the same analysis.] For small i, ΔG_i increases as more atoms of X are added, and so clusters of X form and decay. However there is a critical cluster size above which the addition of further atoms causes ΔG_i to decrease: above this critical size the clusters become more stable as they grow. It is with this critical cluster size that we identify the condensation nucleus.

The size of the nucleus is given by determining the value of ΔG_i which is a maximum with respect to cluster size; in other words we wish to maximize ΔG_i [Eq. (5.4)] with respect to a. Differentiating Eq. (5.4) and equating to zero in the usual way we get the size of a critical cluster to be

$$a^* = \frac{2\gamma m_\mathrm{X}}{\rho kT \log_\mathrm{e} \mathcal{S}} \tag{5.5}$$

and the critical value of ΔG_i^* is

$$\Delta G_\mathrm{i}^* = \frac{16\pi\gamma^3}{3(kT\rho/m_\mathrm{X})^2(\log_\mathrm{e} \mathcal{S})^2}. \tag{5.6}$$

From Eqs. (5.3) and (5.6) the number per unit volume of critical clusters is

$$n_{\rm i}^* = n_1 \exp[-\Delta G_{\rm i}^*/kT]. \tag{5.7}$$

As a matter of interest we can use Eq. (5.5) to get some idea of the size of a nucleation site; the relevant parameters are given in Table 5.1. Inserting some typical values for carbon in Eq. (5.5), taking $\mathcal{S} = 10$ and noting that carbon condenses at $T = 1800\,\rm K$, suggests that a typical nucleation site will have dimensions of a few Ångstroms.

The *rate* at which homogeneous nucleation will take place is given by

$$J_{\rm hom} = 4\pi[a^*]^2 S \left[\frac{P_{\rm X}}{(2\pi m_{\rm X} kT)^{1/2}}\right] Z n_{\rm X} \exp\left[-\frac{\Delta G_{\rm i}^*}{kT}\right], \tag{5.8}$$

where S is the probability that an incident atom X will stick to the cluster and the factor $P_{\rm X}/(2\pi m_{\rm X} kT)^{1/2}$ is the rate at which atoms strike the cluster surface; $P_{\rm X}$ and $n_{\rm X}$ are respectively the pressure and number density of X in the gas phase. Z is a factor that takes account of 'non-equilibrium factors', such as the possible depletion of X from the gas phase as a result of cluster formation and grain growth.

Again we take the example of carbon condensing at $T = 1800\,\rm K$ in the environment of a star where $n_{\rm C}$ might be $\sim 10^3\,\rm m^{-3}$. From Eq. (5.6) we get $\Delta G_{\rm i}^* \simeq 4.2 \times 10^{-19}\,\rm J$, so that Eq. (5.8) leads to $J_{\rm hom} \simeq 10^{-17}\,\rm m^{-3}s^{-1}$. This is so slow that it seems very unlikely that homogeneous nucleation, at least of carbon, is relevant in stellar environments. On the other hand laboratory experiments suggest that silicon carbide, for which there is ample evidence in the environments of stars (see Chapter 7), does nucleate homogeneously, and may even provide the nucleation sites for *heterogeneous* nucleation of carbon in carbon-rich stars.

The difficulties posed by homogeneous nucleation in astrophysical environments has prompted many to look at the possibility that nucleation might occur heterogeneously. Heterogeneous nucleation occurs if the nucleation site differs chemically from the accreting species. For example, nucleation may occur on ions, or on pre-existing molecules such as CO or hydrocarbons. However, in the case of cool carbon stars at least, it seems unlikely that ion nucleation is a significant process because (i) the stars are too cool to ionize any but the elements (like sodium) of lowest ionization potential and (ii) the abundance of these elements are too low to account for the often prolific dust formation seen in these stars. Ion nucleation may however be more important in the atmospheres of hotter stars which are observed to produce dust. Another possibility is that one species may nucleate homogeneously and give rise to a small (dimensions $\simeq$ a few 10's of Ångstroms) grain, which then serves as a nucleation site for the growth of an entirely different species (as in the case of SiC, noted above). Indeed, a grain material such as olivine (Mg_2SiO_4) *must* nucleate heterogeneously: it will obviously not make the transition from the gas phase to the solid phase in a straightforward way, even if only for the fact that we know that Mg_2SiO_4 does not occur in the gas phase.

5.4 Grain condensation temperatures

The condensation temperature $T_{\rm con}$ for a grain species is given by the condition that the vapour pressure $P_{\rm vap}$, given by Eq. (5.1) or (5.2), be equal to the partial pressure of the corresponding condensing species in the gas phase. Again the situation is more straightforward for an elemental condensate (like carbon) than it is for a compound condensate (like magnesium silicate). However even in the simplest of cases there are complications: for example, the details of the chemistry by which nucleation sites form in carbon-rich environments (e.g. whether C_2 or C_2H_2 molecules act as nucleation sites) affects the partial pressure of carbon and hence the value of the condensation temperature. Differences in individual environments (e.g. partial pressure of condensate) will therefore result in corresponding differences in condensation temperature.

Neglecting small particle effects, the condensation temperature for a material X, $T_{\rm con}$ is determined by the condition

$$P_{\rm vap} = P_0 \exp[-T_0/T_{\rm con}] = n_{\rm X} k T_{\rm con}. \tag{5.9}$$

The condensation temperature of several solids of astrophysical interest are given in Table 5.2. For the oxygen-rich condensates ($MgSiO_3$ etc.) the data are for a gas having solar elemental abundances in conditions appropriate for the early solar system; values for other environments are not likely to be substantially different. For the carbon-rich condensates C and SiC the condensation temperatures are those appropriate for the atmospheres of carbon stars. In many cases, where grain formation can be seen in real time in some stellar winds, the observed condensation temperature can be a helpful (if not conclusive) pointer to the nature of the condensate.

5.5 Grain growth

5.5.1 Rate of grain growth

Once nucleation has occurred growth is relatively straightforward. We take the case of a small spherical uncharged grain of material X, with initial radius

Table 5.2: Condensation temperatures for common astronomical solids. Based on L. Grossman, *Geochimica Cosmochimica Acta*, Vol. **36**, 597 (1972)

Condensate	Condensation temperature (K)
H_2O	150–200
Carbon	1850
Silicon Carbide	1500
Fe_3O_4	400
Fe_3C	1100
Al_2O_3	1760
$MgSiO_3$	1350
Mg_2SiO_4	1440
Iron	1470

a_0, immersed in a gas at temperature $T_{\rm gas}$ containing number density $n_{\rm X}$ of atoms of X.

An order-of-magnitude estimate of the rate at which the atoms strike the grain surface may be made if we suppose that all atoms move with a speed corresponding to the mean kinetic energy for atoms in a gas at temperature $T_{\rm gas}$ [cf. Eq. (4.4)]; this gives the rate at which atoms strike the surface to be

$$\mathcal{R} \simeq n_{\rm X}\sigma \left(\frac{3kT_{\rm gas}}{2m_{\rm X}}\right)^{1/2}.$$

More rigorously we proceed in the usual way by determining the strike rate for atoms with velocities in the range $v \rightarrow v + dv$. In a gas of species X, the number of atoms having velocity in this range is, according to Eq. (E.1), $n_{\rm X}f(v)dv$. For a spherical, uncharged grain of radius a, the rate ($\rm s^{-1}$) at which the atoms X strike the grain surface is $\pi a^2 n_{\rm X} v f(v)dv$. Thus the rate at which all X atoms strike the surface, irrespective of velocity, is given by integrating over all velocities:

$$\mathcal{R} = \pi a^2 n_{\rm X} \int_0^\infty v f(v) dv. \tag{5.10}$$

The integral is easily solved by making the substitution $x = m_{\rm X}v^2/2kT_{\rm gas}$ as before, and we have

$$\mathcal{R} = 4\pi a^2 n_{\rm X} \left(\frac{kT_{\rm gas}}{2\pi m_{\rm X}}\right)^{1/2}. \tag{5.11}$$

If the 'sticking probability' is S, the rate at which atoms stick to the surface is

$$\mathcal{R}' = 4\pi a^2 n_{\rm X} S \left(\frac{kT_{\rm gas}}{2\pi m_{\rm X}}\right)^{1/2} \tag{5.12}$$

and the rate at which the grain mass increases is

$$\dot{m}_{\rm grain} = m_{\rm X}\mathcal{R}' = 4\pi a^2 n_{\rm X} m_{\rm X} S \left(\frac{kT_{\rm gas}}{2\pi m_{\rm X}}\right)^{1/2}. \tag{5.13}$$

However, if we suppose that the atoms strike and stick to the surface uniformly all over the (spherical) surface, we can also express the rate of increase of grain mass as

$$\dot{m}_{\rm grain} = 4\pi a^2 \rho \dot{a}, \tag{5.14}$$

where ρ is the density of the grain material. Equating (5.13) and (5.14) we get

$$\dot{a} = \frac{n_{\rm X} S}{\rho}\left(\frac{kT_{\rm gas} m_{\rm X}}{2\pi}\right)^{1/2}. \tag{5.15}$$

Eq. (5.15) enables us to determine the dependence of grain radius on time, although we have to bear in mind that $n_{\rm X}$, the number density of X in the gas phase, may also depend on time. This would happen if, for example, grain growth depleted the gas-phase abundance of X to such an extent that $n_{\rm X}$ decreased significantly as a result. Alternatively the grains may be growing in a stellar wind, in which case $n_{\rm X}$ will decrease as a result of the thinning out of the outflow (see Chapter 7).

We can easily integrate Eq. (5.15) if $n_{\rm X}$ and $T_{\rm gas}$ do not depend on time; we get

$$a(t) = a_0 + \frac{n_{\rm X} S}{\rho}\left(\frac{kT_{\rm gas} m_{\rm X}}{2\pi}\right)^{1/2} t, \tag{5.16}$$

where a_0 is the size of the nucleation seed. We see that the grain grows linearly with time and, if there is ample supply of the condensate (i.e. if $n_{\rm X}$ is large enough), there is no limit to the size the grain eventually attains. Eventually however the growth of grains will result in a significant decrease of $n_{\rm X}$: *depletion* of the gas-phase species will become important.

5.5.2 The effect of depletion of the condensing species

The above discussion therefore implicitly assumes that there is an infinite supply of atoms in the gas phase, so that the value of $n_{\rm X}$ does not decrease as atoms condense out of the gas phase onto the surface of the grains: the depletion of the gas is ignored. We can take depletion into account in a slight modification of the above analysis, as follows. The rate at which atoms stick to the surface of a spherical grain of radius a is given by the $\mathcal{R}'$ of Eq. (5.12); this also represents the rate at which gas atoms are removed from the gas phase. If there are $n_{\rm d}$ dust grains of radius a per unit volume, then $dn_{\rm X}$ atoms are removed from the gas phase in time dt, where

$$\frac{dn_{\rm X}}{dt} = -4\pi a^2(t) n_{\rm X}(t) n_{\rm d} S \left(\frac{kT_{\rm gas}}{2\pi m_{\rm X}}\right)^{1/2} ; \tag{5.17}$$

note that, in this case, *both* a and $n_{\rm X}$ are time-dependent.

The rate at which the mass of a grain of radius a increases is again given by Eq. (5.13), except that again, $n_{\rm X}$ is time-dependent:

$$\dot{m}_{\rm grain} = 4\pi a^2 n_{\rm X}(t) m_{\rm X} S \left(\frac{kT_{\rm gas}}{2\pi m_{\rm X}}\right)^{1/2} . \tag{5.18}$$

Consequently Eq. (5.15) still applies, provided that we remember the time-dependence of $n_{\rm X}$. We therefore have a set of simultaneous differential equations describing the variations of a and $n_{\rm X}$ with time:

$$\frac{dn_{\rm X}}{dt} = -4\pi a^2(t) n_{\rm X}(t) n_{\rm d} S \left(\frac{kT_{\rm gas}}{2\pi m_{\rm X}}\right)^{1/2} \tag{5.19}$$

$$\frac{da}{dt} = \frac{n_{\rm X}(t) S}{\rho} m_{\rm X} \left(\frac{kT_{\rm gas}}{2\pi m_{\rm X}}\right)^{1/2} . \tag{5.20}$$

We proceed by differentiating Eq. (5.19) with respect to time and substituting for $dn_{\rm X}/dt$ from Eq. (5.20); the result is, after further substituting for $n_{\rm X}$:

$$\ddot{a} = \frac{d^2 a}{dt^2} = -4\pi a^2 \overline{v}_{\rm X} S n_{\rm d} \dot{a}, \tag{5.21}$$

where we have replaced the expression $\{(kT_{\rm gas})/(2\pi m_{\rm X})\}^{1/2}$, which just represents the thermal velocity of the gas atoms, by $\overline{v}_{\rm X}$ for brevity. We expect that, as the condensing species becomes depleted from the gas phase, the grain radius will attain a final value a_∞ and the rate of grain growth will have become zero. Eq. (5.21) is therefore easily integrated by writing

$$\frac{d^2 a}{dt^2} = \dot{a}\frac{d\dot{a}}{da};$$

the result is

$$\dot{a} = \frac{4\pi \overline{v}_{\rm X} S n_{\rm d}}{3}(a_\infty^3 - a^3). \tag{5.22}$$

Integrating Eq. (5.22) does not, unfortunately, give a (and hence $n_{\rm X}$) explicitly in terms of t. What we can estimate, however, is the final size of

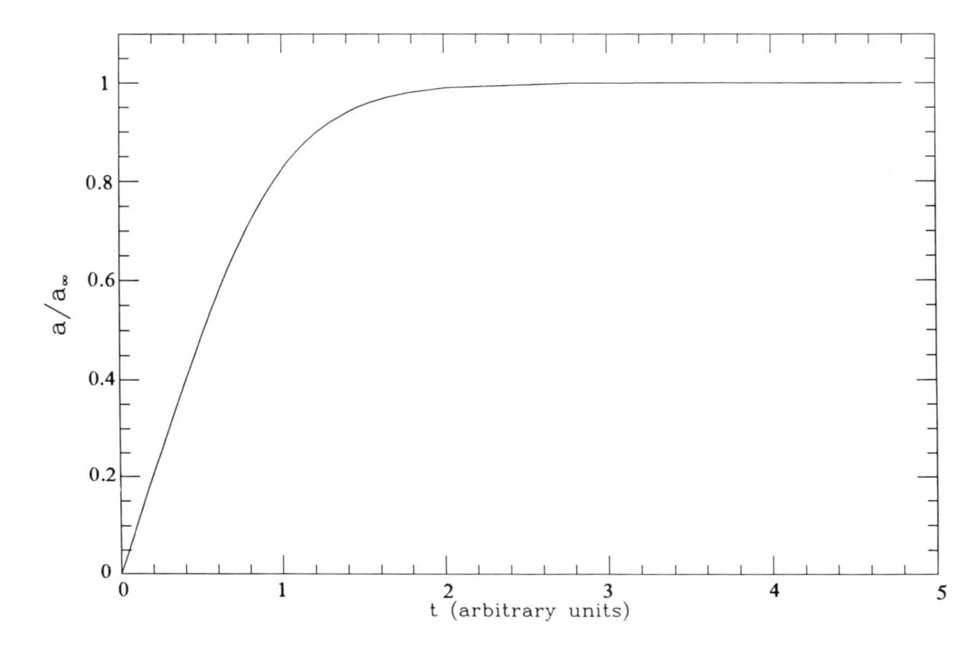

Figure 5.2: The dependence of grain size on time when the condensing species is depleted.

the grains a_∞. During the earliest stages of the grain growth, we suppose that $a_\infty \gg a_c$, the size of a condensation nucleus, so that around $t = 0$, $\dot{a}(t = 0) \simeq 4\pi \overline{v}_X S n_d a_\infty^3/3$ from Eq. (5.22). Now during this time the rate of growth is likely to be such that the gas is not noticeably depleted, so that $\dot{a}$ may also be approximated by Eq. (5.15). Equating these two estimates of $\dot{a}$ and solving for a_∞ we find

$$a_\infty \simeq \left(\frac{3m_X n_X(t=0)}{4\pi \rho n_d}\right)^{1/3}.$$

Note that, the greater the number of dust grains per unit volume n_d, the smaller the final size of the grain; this is what we would expect intuitively as there are fewer condensing atoms per grain in this case: the available condensate has to be shared between more grains. Also the larger the initial $n_X(t = 0)$ concentration of condensing species, the larger the final size of the grain, again as we would expect intuitively.

Fig. 5.2 shows the dependence of a on time as determined by integrating Eq. (5.22). The eventual levelling off of grain growth is evident, as is the asymptotic approach to the final grain size a_∞; note also the initial linear rise while the condensing vapour is still plentiful. We can get an estimate of the time τ_∞ it takes for the grain to attain maximum size from $\tau_\infty \simeq a_\infty/\dot{a}(t=0)$, i.e. the time it would take for the grain to reach a_∞ had the initial linear rise been maintained. We find

$$\tau_\infty \simeq \frac{3}{4\pi \bar{v}_X S n_d a_\infty^2} \simeq \left(\frac{3m_X\rho^2}{4\pi}\right)^{1/3} \frac{1}{S}\left(\frac{2\pi}{kT_{gas}m_X}\right)^{1/2} \frac{1}{[n_d n_X^2(t=0)]^{1/3}} \tag{5.23}$$

5.5.3 Nature of grain

In the previous subsection we looked at the rate at which a grain grows. We now consider: will the resulting grain be amorphous or crystalline? In order to determine this we need to look beyond '...the rate at which atoms stick to the surface...' and consider *how* the atom or molecule binds to the grain surface. The surface will consist of a number of potential wells of varying depth and, given time, we can expect an impinging atom or molecule to migrate over the grain surface and bind to the surface at the energetically most favourable site; in such a case the resulting grain will be crystalline. However if the rate of arrival at of atoms at the surface is such that there is no time for each individual atom or molecule to find the most favourable site the atom will bind where it can and an amorphous grain will result.

Whether the resulting grain is crystalline or amorphous will therefore be determined, at least in part, by the relative magnitudes of the surface migration time τ_s and the time interval τ_i between arrivals of atoms from the vapour phase onto the grain surface. For a grain of radius a immersed in a gas at pressure P and temperature T_{gas} the time interval

$$\tau_i = \left(\frac{P}{2\pi mkT_{gas}}\right)^{1/2} \pi a^2 S, \tag{5.24}$$

where S is the sticking probability. The migration time is [cf. Eq. (4.24)]

$$\tau_m = \nu_0^{-1} \exp\left[\frac{E_A}{kT_d}\right], \tag{5.25}$$

where the activation energy E_A is about 2.9 eV for carbon.

If $\tau_m < \tau_i$ then there is ample time for the incident atom to migrate over the grain surface and 'stick' at the most favourable site, leading to the growth of a crystalline grain. However if $\tau_m > \tau_i$ then impinging atoms have no time to find the most favourable site before the next atom arrives. Obviously whether or not a grain is crystalline will depend on the grain and gas temperatures and on the activation energy of the grain material.

5.6 Grain destruction

There are several mechanisms whereby dust grains are destroyed in a variety of astrophysical environments. The main destructive agents include evaporation and sputtering, but indirect evidence suggests that grain-grain collisions are also important.

5.6.1 Evaporation

Suppose we have a spherical grain, of radius a and temperature T_d, in a vacuum. The grain will lose atoms from its surface as a result of evaporation, and this process will continue until there is equilibrium between the loss of atoms by evaporation and recondensation onto the surface from the vapour phase. During the evaporation process the grain radius decreases and this is given by the equivalent of Eq. (5.15):

$$\dot{a} = -\frac{P_\mathrm{vap}(T_\mathrm{d})}{\rho}\left(\frac{m_\mathrm{X}}{2\pi k T_\mathrm{d}}\right)^{1/2}, \tag{5.26}$$

with P_vap given by Eq. (5.1). If the temperature remains constant while the grain evaporates, we can immediately integrate Eq. (5.26) to determine the dependence of grain radius on time:

$$a(t) = a_0 - \frac{P_\mathrm{vap}}{\rho}\left(\frac{m_\mathrm{X}}{2\pi k T_\mathrm{d}}\right)^{1/2} t, \tag{5.27}$$

where a_0 is the initial radius of the grain. The grain radius reaches zero–i.e. a grain of initial radius a_0 evaporates away completely–in a time given by

$$\tau_\mathrm{evap} = \frac{a_0\rho}{P_\mathrm{vap}}\left(\frac{2\pi k T_\mathrm{d}}{m_\mathrm{X}}\right)^{1/2}. \tag{5.28}$$

We have already seen, however, that under certain circumstances, the grain temperature depends on grain radius a; in this case Eq. (5.26) cannot be integrated analytically. However provided that $T_0 \gg T_\mathrm{d}$, which is always the case, and $T_\mathrm{d} \propto a^{-1/(\alpha+4)}$ [see Eq. (4.22)], we can write

$$\tau_\mathrm{evap} \simeq \frac{a_0\rho}{P_\mathrm{vap}}\left(\frac{2\pi k T_\mathrm{d}}{m_\mathrm{X}}\right)^{1/2} (\alpha+4)\exp[T_\mathrm{d}/T_0](T_\mathrm{d}/T_0), \tag{5.29}$$

which is a factor $(\alpha+4)\exp[T_\mathrm{d}/T_0](T_\mathrm{d}/T_0)$ less than that given by Eq. (5.28). The reason for this difference is easy to see: if the grain temperature depends on a as in Eq. (4.22), the grain gets hotter as it evaporates and the evaporation will accelerate. We can therefore expect the evaporation time to be considerably smaller in this case.

5.6.2 Sputtering

If an atom or ion strikes the surface of a solid with sufficient momentum, it is likely to eject atoms from the surface of the solid with the result that the grain is eroded. This process is known as sputtering, and is well-known in terrestrial environments such as furnaces. Although the physics of sputtering is still not well understood in detail, it seems that sputtering consists of a number of sub-processes, such as the displacement of an atom from a lattice site, the diffusion of the ejected atom through the surface and ejection from the surface.

At the simplest level, we can regard sputtering as a simple momentum transfer process, in which case the maximum energy transfer ΔE for an incident projectile of energy E is

$$\Delta E = \frac{4m_{\rm p}m_{\rm t}}{(m_{\rm p}+m_{\rm t})^2}E = \eta E,$$

where $m_{\rm p}$ and $m_{\rm t}$ are the atomic masses of the 'projectile' and 'target' atom respectively. If the binding energy of an atom close to the surface of the solid is U_0, then no sputtering can occur if $\eta E < U_0$; in other words there is a threshold energy below which sputtering can not take place.

Laboratory work on sputtering suggests that it is useful to define a quantity called the *sputtering yield* $Y_{\rm s}$. Suppose we have a flux of (say) protons incident on a carbon surface; the yield tells us how many carbon atoms are ejected per incident proton. The yield actually depends on a number of factors, such as the nature and energy of the bombarding particles, the nature of the surface, the angle at which the particles strike the surface etc.; indeed it is complications such as these that make sputtering so difficult to understand. Nonetheless typical values for $Y_{\rm s}$ are about 0.01 for 5 keV protons incident on graphite.

We can determine the rate at which a spherical grain, radius a, is eroded by sputtering in much the same way as we determined the growth rate in Section 5.5. Indeed we might simply take Eq. (5.12) and just replace the sticking probability S by the sputtering yield $Y_{\rm s}$ to obtain the rate at which atoms are removed from the grain surface. However this approach does not take into account the fact that sputtering has a threshold, U_0/η; the actual rate at which surface atoms are removed is therefore

$$\mathcal{R}'' = -\pi a^2 n_{\rm p} Y_{\rm s} \int_{v_0}^{\infty} v f(v) dv, \tag{5.30}$$

where $\frac{1}{2}m_{\rm p}v_0^2$ is the kinetic energy corresponding to the threshold energy and $n_{\rm p}$ is the number of gas-phase (projectile) atoms per unit volume. The integral in Eq. (5.30) is again easily solved by making the substitution $x = m_\mathrm{p}v^2/2kT_\mathrm{gas}$; the result is

$$\mathcal{R}'' = -4\pi a^2 n_\mathrm{p} Y_\mathrm{s} \left(\frac{kT_\mathrm{gas}}{2\pi m_\mathrm{p}}\right)^{1/2} (1 + x_0)\exp[-x_0], \tag{5.31}$$

where $x_0 = U_0/\eta kT_\mathrm{gas}$.

We should further note that the bombarding atoms or ions are not (as they were in the case of the grain growth problem) necessarily the same as the atoms from which the grain is made, i.e. $m_\mathrm{p} \neq m_\mathrm{t}$. As we have already noted, sputtering seems to be a momentum transfer process so we would expect the more massive atoms to be more effective at eroding the grain away. While this expectation seems to be borne out by laboratory experiments we must remember that, in general, the more massive an atom the lower its abundance in a cosmic environment. Thus despite their low mass, it is generally bombardment by protons (and to a lesser extent, helium) that does the damage to interstellar grains.

The rate at which the grain mass *decreases* is, therefore, by analogy with Eq. (5.13), given by

$$\dot{m}_\mathrm{grain} = -4\pi a^2 n_\mathrm{p} m_\mathrm{t} Y_\mathrm{s} \left(\frac{kT_\mathrm{gas}}{2\pi m_\mathrm{p}}\right)^{1/2} (1 + x_0)\exp[-x_0], \tag{5.32}$$

where m_t is the atomic or molecular mass of the (target) material from which the grain is made. Using Eq. (5.14) we get for the rate at which a spherical grain is eroded

$$\dot{a} = -\frac{n_\mathrm{p} m_\mathrm{t} Y_\mathrm{s}}{\rho} \left(\frac{kT_\mathrm{gas}}{2\pi m_\mathrm{p}}\right)^{1/2} (1 + x_0)\exp[-x_0]. \tag{5.33}$$

A more rigorous treatment would take into account the possibility that the grain might be charged, the dependence of Y_s on energy and the fact that, in a gas having cosmic abundances, hydrogen, helium and (to a lesser extent) heavier elements would also be effective (and would also of course have different dependences of Y_s on energy).

5.6.3 Chemical sputtering

Like the sputtering process discussed in the previous subsection, chemical sputtering (or chemisputtering) can cause grain erosion. In this case however the sputtering takes place via a series of chemical reactions on the grain surface. For example, the surface reaction

$$H_2O + H \rightarrow OH + H_2$$

may erode away an ice grain by virtue of the bombardment of the grain by protons, while the bombardment of a carbon surface by protons may remove carbon atoms, in the form of hydrocarbons, from the surface of a graphite grain.

The derivation of the rate at which a grain is eroded by chemisputtering is similar to the determination of the erosion by physical sputtering just discussed. However there are three factors that are different in the chemisputtering case. First, unlike the sputtering process discussed in the previous section, there is no threshold energy for chemisputtering. Second, the chemisputtering process is fairly specific in terms of the bombarding species and the grain material, for example carbon dust can be chemically eroded by hydrogen bombardment, but not by helium. Third, not only must we (as we did in the case of ordinary sputtering) take account of the chemisputtering yield Y_c, we must also remember that it is going to take four hydrogens to lift a carbon atom off the grain surface. We therefore have

$$\dot{a} = -\frac{n_H m_C Y_c}{4\rho}\left(\frac{kT_{gas}}{2\pi m_H}\right)^{1/2} \tag{5.34}$$

for the erosion of a carbon grain by hydrogen.

5.6.4 Grain-grain collisions

There are circumstances in which the two dust grains will collide and the outcome of the collision can depend on the circumstances. If the total energy in the centre-of-mass reference frame exceeds the vaporization energy then the result of the collision is that both grains will be vaporized.

Another possibility is that the two grains are shattered, leaving smaller fragments. Laboratory experiments and simple theory (see Appendix B) suggest that when this happens, the mass distribution of fragments (i.e. the number of fragments with mass m in the range $m \rightarrow m + dm$) $n(m)$ is given by

$$n(m)dm \propto m^{-k} dm, \tag{5.35}$$

where k is a constant. For example, in the Solar System there is a region between the orbits of Mars and Jupiter, the asteroid belt, in which there are several thousand bodies ranging in size from ~ 1000 km downwards; for the asteroid population $k \simeq 1.8$. In the context of interstellar dust grains, however, it is more convenient to consider the *size* distribution, i.e. the number $n(a)da$ of grains with radius a in the range $a \to a + da$. Now

$$n(a)da = n(m)dm,$$

where the size interval da corresponds to the mass interval dm. Since

$$m = \frac{4\pi a^3 \rho}{3}$$

for spherical grains, ρ being the density of the grain material, we have that

$$dm = 4\pi a^2 \rho da.$$

Therefore

$$n(a) = n(m)\frac{da}{dm} \propto m^{-k} a^2 \propto a^{-l}, \tag{5.36}$$

where $l = 2 - 3k$: a power-law distribution of grain masses also represents a power-law distribution of grain radii; thus in the asteroid belt the size (rather than mass) distribution has exponent $l = 3.4$. We shall see in Chapter 6 that a similar size distribution may be inferred for interstellar dust grains. As is shown in Appendix B, such a grain size distribution can arise as a result of grain-grain shattering, a simple treatment of which leads to a size distribution

$$n(a)da \propto a^{-3} da. \tag{5.37}$$

Even though the analysis that leads to Eq. (5.37) is rather simplistic it suggests that the size distribution of particles in the Solar System and beyond may be governed to a large extent by the shattering of larger bodies.

5.6.5 Other grain destruction mechanisms

We have been concerned mainly with spherical grains, largely because spherical symmetry is easy to treat analytically. However we can expect that, in general, grains need not be spherical. Indeed, the polarization of starlight discussed in Chapter 6 is strong evidence that interstellar grains (at least those responsible for polarization) are certainly *not* spherical. In the extreme case of a highly elongated grain, or a spherical grain with pronounced 'spikes', the electric field at the ends of a charged, elongated grain, or at the 'spike' of a charged spherical grain, may be sufficiently high that positively charged ions are ejected from the surface by the process of field ion emission. Indeed we might expect that grains having sharp 'spikes' have their spikes removed by this process.

A further difficulty that charged grains may suffer is that, if the charge is sufficiently high, the resultant electrostatic tensile stress may exceed the tensile strength of the grain material. We consider the case of a spherical grain on whose surface there is charge Ze (cf. Section 4.2); the surface density of charge, in coulomb m^{-2}, is $\sigma = Ze/4\pi a^2$. As is well-known, the presence of the charge results in an outward force per unit area (i.e. a stress), irrespective of the sign of the charge on the surface. The magnitude of this stress is

$$F = \frac{\sigma^2}{2\epsilon_0}, \tag{5.38}$$

and if this exceeds the tensile strength of the grain material then disruption of the grain will result. For a 0.1 μm grain carrying a charge given by $Z \simeq 100$, the resultant stress is $\sim 10^4$ N m^{-2}. This value is typical of weakly-bound solids, such as might be found in (for example) a cometary environment or in porous aggregates of smaller grains which are also found in the solar system.

Problems

5.1. A 0.1 μm carbon grain (for which $U_0 = 7.4$ eV) is immersed in hydrogen gas having $n_{\rm H} = 10^6$ m^{-3} and temperature (i) 10^4 K, (ii) 10^5 K. What are the timescales in each case for the complete erosion of the grain by (i) physical sputtering, (ii) chemisputtering? Assume that the sputtering yield is 10^{-2} for the former process and 10^{-4} for the latter.

5.2. A spherical asteroid, of initial radius 10 km, is injected into a circular orbit of radius 0.03 AU. Assuming that the asteroid is made of silicate, how many orbits of the Sun can it make before it is completely evaporated? (Assume that the both the Sun and the asteroid behave like blackbodies for the purposes of calculating the asteroid temperature).

5.3. Derive the equivalent of Eq. (5.34) for the erosion of ice grains by proton bombardment.

Reading

The following is a good undergraduate text on thermodynamics and statistical physics:

[C] F. Mandl, *Statistical Physics*, Wiley (1988).

The paper by Draine gives a comprehensive review of grain nucleation and growth:

[D] B. T. Draine, *Astrophysics & Space Science*, Vol. **65**, 313 (1979).

while the following paper provides a very full discussion of destruction mechanisms for interstellar grains:

[D] B. T. Draine & E. E. Salpeter, *Astrophysical Journal*, Vol. **231**, 438 (1979).

6

Interstellar dust

6.1 Introduction

In discussing interstellar dust, we should distinguish between dust in the general diffuse interstellar medium, and dust in denser regions (interstellar clouds), as it transpires that dust grains residing in dense clouds have features that are different from those in the general medium. We have already mentioned that grains in dense clouds play a vital rôle in the chemistry of clouds. Also, in many cases grains in dense clouds may have acquired a mantle of ice, so that the average dimensions of grains in clouds are larger than those in the diffuse medium. In this chapter we look at the properties of dust grains in interstellar space.

6.2 The extinction law

One of the most obvious properties of interstellar dust that can be determined observationally is the extinction law, i.e. the way in which the extinction properties of the dust depend on wavelength. This has been particularly fruitful in recent years, with the extension of observing capabilities beyond the optical, into the ultraviolet and infrared. Before discussing the results of this work, we should first consider how the extinction law is derived from astronomical observations.

6.2.1 The extinction

Suppose we have a source which would have a flux S_0 in the absence of any interstellar extinction. How does the observed flux depend on the way in which the intervening medium scatters and absorbs radiation?

We have already laid the foundations for this in Chapter 2. In the case of radiation emitted by a star and propagating through the interstellar medium, we can neglect emission by the medium itself (at least at optical wavelengths) so that $\mathcal{S}_\nu = 0$ in Eq. (2.7). Let us now integrate Eq. (2.6) over a path length D; after re-arranging we get

$$\int_0^D \frac{dS}{S} = -\kappa_\nu dx \tag{6.1}$$

so that

$$S = S_0 \exp\left[-\int_0^D \kappa_\nu dx\right]. \tag{6.2}$$

Now the integral in Eq. (6.2) is of course the optical depth τ_ν, as defined in Eq. (2.3); obviously we have $\tau_\nu = \kappa_\nu D$ if the extinction coefficient is independent of position.

We can therefore determine the extinction law in the direction of a star if we measure its flux distribution S_λ, and compare this with the flux distribution S_λ^0 the star *would* have had in the absence of any intervening material. In practice this means comparing S_λ with the flux distribution of a star which is known, from other considerations, to be unaffected by the interstellar medium.

The extinction coefficient κ_ν can be expressed as the product of the number n of absorbers/scatterers per unit volume, and their extinction cross-section σ_{ext}. We can also write $\tau_\nu = \int n\sigma_{\text{ext}}dx = \sigma_{\text{ext}} \int ndx$ if the agents responsible for the extinction are identical. The integral $\int ndx$ is commonly known as the *column density*.

6.2.2 The extinction in magnitudes

Although the above provides us with the means of determining τ_λ, actually doing this at high wavelength resolution can be extremely time-consuming. It is often more convenient to obtain, at least initially, an extinction law with coarse wavelength resolution using the broad-band filters described in Chapter 2.

Converting the optical depth τ_λ to an extinction in magnitudes is straightforward. We can use Eq. (6.2), together with Eq. (1.9), to convert the flux to magnitudes. First take logs to base 10 of both sides of Eq. (6.2) and then multiply both sides by -2.5:

$$-2.5\log_{10} S = -2.5\log_{10} S_0 + 2.5\log_{10}[\kappa D] \tag{6.3}$$

or

$$m = m_0 + \frac{2.5}{\log_e 10}\tau = m_0 + A, \tag{6.4}$$

where m_0 is the apparent magnitude one would observe in the absence of extinction and A is the extinction in magnitudes. Eq. (1.17) is therefore more completely written as

$$m = M + 5\log_{10} D + A - 5. \tag{6.5}$$

Note that the effect of extinction is to make the object appear fainter (larger m) and therefore more distant, than it would in the absence of extinction. Specifically we can define the extinction A_V for the visual magnitude V, A_B for the blue magnitude B, and so on, for all the magnitudes described in Chapter 2.

6.2.3 Colour excess

As well as the extinction at individual wavelengths, it is also useful to know the colour excess in the direction of a particular object; this arises as follows. Suppose we observe a star, the spectral type of which is known; the spectral type determines its intrinsic colour indices $(B-V), (U-B), (J-H)$ and so on (see Section 1.3.3). Suppose we now measure the colour indices of the star. We will in general find that the measured colour indices are greater than the indices expected on the basis of spectral type and the star is said to have a *colour excess*, generally denoted by E. Thus for example, if we denote the expected colour index by $(B-V)_0$ and the corresponding observed index by $(B-V)$ then

$$E(B-V) = (B-V) - (B-V)_0 \tag{6.6}$$

$$= A_B - A_V. \tag{6.7}$$

Obviously there are equivalent definitions for $E(J-H)$ etc. Since the value of E is invariably positive, irrespective of which wavelength pair is chosen, the star is said to be *redder* than it would be in the absence of any intervening medium. Thus the effect of interstellar (and indeed circumstellar) dust is not only to extinguish starlight but also to redden it: in general we have both extinction and reddening.

There is one possible complication that has to be borne in mind in the discussion of interstellar reddening. As discussed in Chapter 7 many stars have circumstellar dust shells and the dust is often hot enough to radiate at near infrared wavelengths–in other words the dust will contribute to the flux at (say) K and L magnitudes. The colour excess determined from infrared colour indices will not then of course give a reliable measure of reddening.

6.2.4 The ratio of total-to-selective extinction

In principle we can then determine the extinction A_λ as a function of wavelength λ. However, all astronomical observation tells us is the observed magnitude [m in Eq. (6.4)]; in order to determine the extinction law we must know the wavelength-dependence of m_0 as well. This can generally be done if we know, from other observations, the way in which m_0 is expected to vary with λ. For example, the dependence of m_0 on wavelength has been determined for a number of nearby stars, having a variety of spectral types and luminosity classes, and which are known (e.g. because they are so close to us) to be unaffected by the presence of intervening material. Thus if we wish to determine the extinction law in the direction of a particular star the spectral type of which is known, all we need do is to compare the wavelength-dependence of A for the star under consideration with that for an otherwise identical star which is known to be unaffected.

A further complication arises in that, while the general form of the extinction law A_λ can be determined, the curve has to be normalized in some way: we must be able to determine the extinction at some particular wavelength so that the level of the entire curve can be determined. However, this is not as straightforward as it seems: how do we determine (for example) the extinction A when all we know in Eq. (6.4) is the observed visual magnitude V? One way is to use the fact that the ratio

$$R = \frac{A_V}{E(B-V)} \tag{6.8}$$

is reasonably constant irrespective of the line of sight chosen through the interstellar dust layer. This quantity is called the *ratio of total-to-selective extinction* as it compares the total extinction (as typified by A_V) with the wavelength-dependence of extinction, as described by the colour excess $E(B-V)$. The value of R turns out to depend on the size of grain responsible for extinction and although a typical value is about 3.1 there are variations, amounting to about 10 per cent, for different lines of sight through the interstellar dust layer. This value of R applies only to the general interstellar dust, larger values being found for dust in denser interstellar clouds and for circumstellar dust.

6.3 The wavelength-dependence of extinction

6.3.1 Extension beyond the optical

However, it is by extending the extinction law to longer and (in particular) to shorter wavelengths that real progress has been made, particularly with the

IUE satellite observatory. The extinction law is basically the dependence of A_λ [or equivalently, of τ_λ], on wavelength. However, in order that extinction laws for different stars, seen through different amounts of interstellar dust, can be directly compared, it is necessary to suitably normalize the various extinction laws. This is done as follows. Instead of looking at A_λ as a function of wavelength, we look at $E(\lambda - V) = A_\lambda - A_V$. This effectively normalizes the extinction law around the wavelength λ_V: all stars will have $E(\lambda - V) = 0$ at this wavelength and the extinction law is normalized with respect to overall extinction. The extinction law is further normalized with respect to the reddening, by dividing by $E(B - V)$. Thus the quantity normally plotted in an extinction law is

$$X = \frac{E(\lambda - V)}{E(B - V)}. \tag{6.9}$$

Also, rather than plot X against wavelength λ, it is more usual to plot the extinction law in terms of λ^{-1}.

A further advantage of plotting the extinction law in this way now emerges. As discussed in Chapter 4, it is expected that the extinction efficiency $Q_{\rm ext} \to 0$ as $\lambda \to \infty$. Now since $A_\lambda \propto \tau_\lambda \propto Q_{\rm ext}$ the extinction will also approach zero at very long wavelengths. Thus the intercept of the extinction law on the ordinate axis is [from Eq. (6.9)] simply

$$X(\lambda \to \infty) = -\frac{A_V}{E(B - V)} = -R. \tag{6.10}$$

In other words the value of R [Eq. (6.8)] may be read directly off as the intercept on the ordinate axis.

We also note here that, in the case of interstellar dust, the colour excess $E(U - B)$ scales linearly with $E(B - V)$:

$$\frac{E(U - B)}{E(B - V)} = 0.72.$$

Consequently the quantity

$$Q = (U - B) - 0.72(B - V)$$

for any star is independent of the reddening in the direction of the star, provided the star suffers no additional circumstellar extinction.

Before we look at extinction laws in general, we look at the ultraviolet (1000–3200Å) spectra of two stars, one of which is known, from its spectral type and $(B - V)$ colour, to be heavily reddened and another (of similar spectral type), which is known to be effectively unreddened (see Fig. 6.1). Both stars are of early B type, the star[1] HD 31726 having spectral type B1V and $(B-V) = -0.21$, and the other (HD 37367) having spectral type B2IV-V and $(B-V) = 0.16$. The intrinsic colour indices are $(B-V)_0 = -0.26$ for HD 31726 and $(B-V)_0 = -0.24$ for HD 37367; note that HD 31726 has $E(B-V) = (B-V) - (B-V)_0 \simeq 0$, consistent with its being unreddened, while HD 37367 has $E(B-V) \simeq 0.40$. There are two things that we see immediately from Fig. 6.1. First, the fact that the flux from the reddened star is increasingly affected as we go to shorter and shorter wavelengths: the 'reddening' that we see in the infrared and optical persists into the ultraviolet. Second, there is a deep, broad extinction feature centred at wavelength $\sim$ 2175Å. Both these features are almost certainly connected with the heavy optical reddening and extinction suffered by HD 37367.

It is now of interest to ask whether the features seen in Fig. 6.1 are common in the ultraviolet spectra of reddened stars and in particular, are the shape and central wavelength of the broad extinction feature the same in all reddened stars? A selection of extinction laws, normalized as described above, are illustrated in Fig. 6.2. One thing that is clear from Fig. 6.2 is that there is indeed a basic similarity between the extinction laws in various lines of sight in the Galaxy: the interstellar grain population is evidently fairly uniform irrespective of location. This is very useful indeed if all we are concerned with is the use of the extinction law to correct the flux distributions of distant objects for the effects of interstellar extinction. However if we try to construct a 'standard' extinction law which might serve for this purpose we find that the extinction curves are not as similar as they at first sight appear. There are, for example, differences in the shape, height and even central wavelength of the '2175Å' bump. Now this could mean that there are real differences–small but significant–in the extinction law in various directions, or it may be that there are subtle differences in the extinction law only around the wavelengths λ_B and λ_V. The latter would result in apparent differences in the overall extinction because the colour excess $E(B-V)$ is used to normalize the entire curve. Nonetheless the underlying similarity in the wavelength-dependence of the interstellar extinction suggests considerable uniformity in the distribution of interstellar dust.

---

[1]HD = Henry Draper catalogue, a catalogue of bright stars compiled at the end of the 19th century.

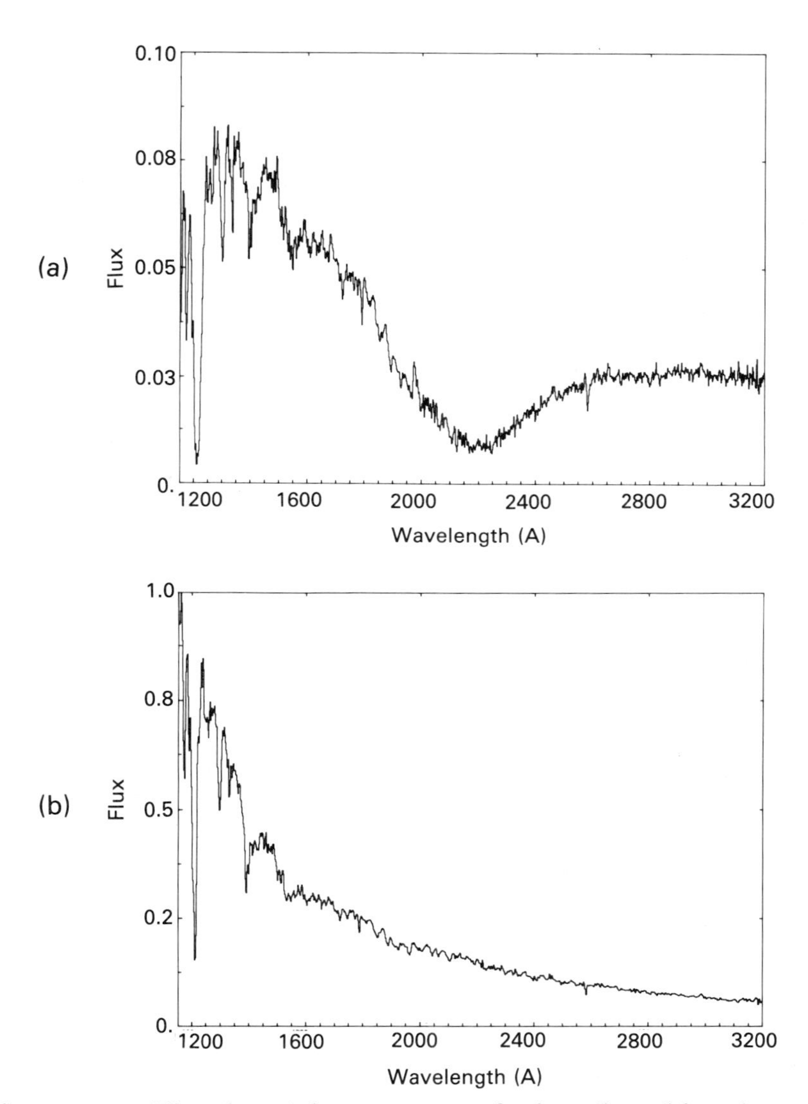

Figure 6.1: The ultraviolet spectrum of a heavily reddened star (top) and an unreddened star of the same spectral type (bottom). Flux density units are $10^{-8}\,\mathrm{W\,m^{-2}\,\mu m^{-1}}$. After A. Heck et al., *IUE low-dispersion spectra reference atlas. Part I. Normal Stars*, European Space Agency Publication ESA-SP-1052 (1984).

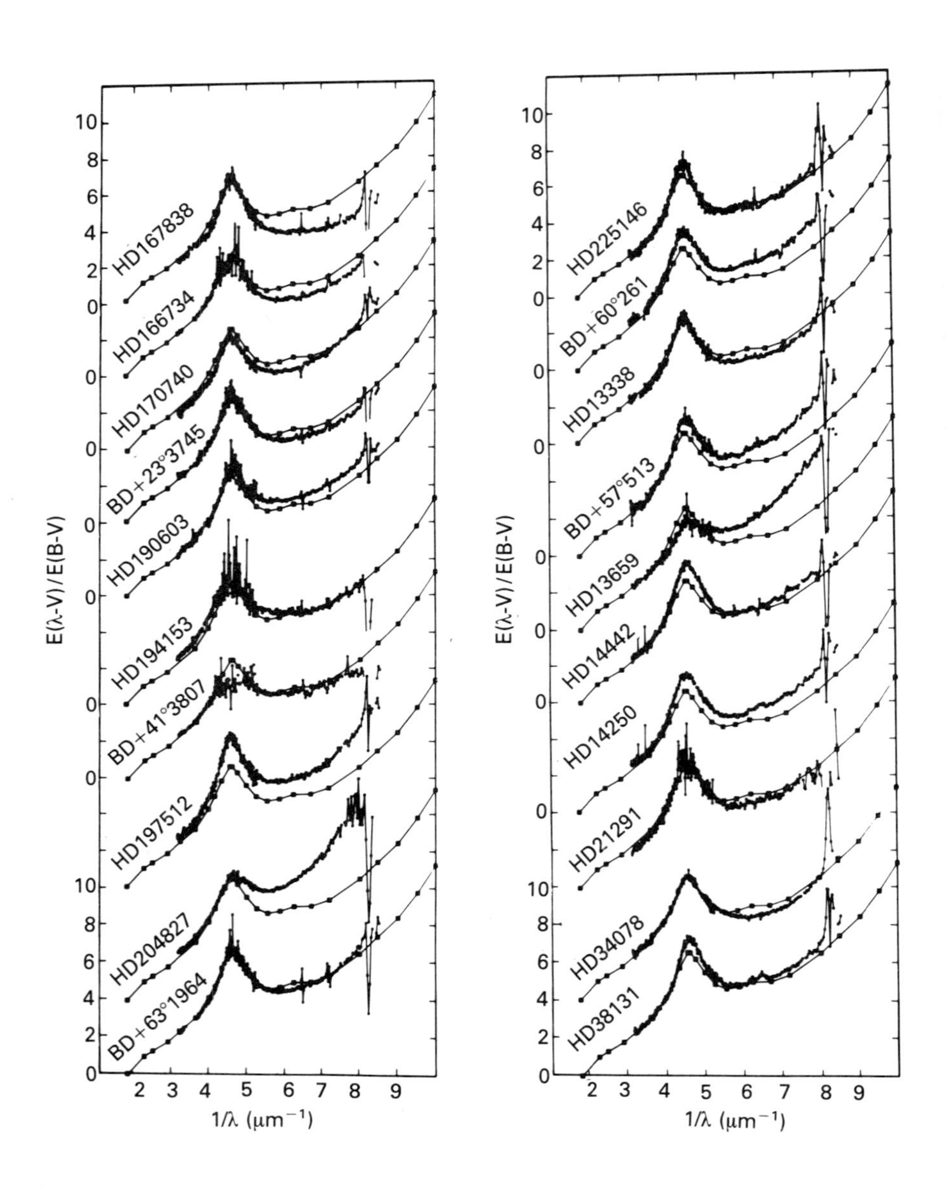

Figure 6.2: Normalized extinction laws for reddened stars. After A. Witt et al., *Astrophysical Journal*, Vol. **279**, 698 (1984).

6.4 Composition of interstellar dust

In discussing the chemical composition of interstellar dust there are several factors that can guide us. First we can ask the question 'how much matter is there in interstellar space?' because clearly the amount of dust in interstellar space must be less than the total amount of matter present. We also have to consider the question 'which elements are dust grains likely to be composed of?' because we know the general cosmic abundance of the elements. Finally are there any dust-related absorption features which are detectable in the spectra of heavily obscured background stars? We consider these points in this section.

6.4.1 Implications for grain size

We first discuss the extinction law, the wavelength-dependence of interstellar extinction as determined from the ultraviolet (~ 1000 Å) to the far infrared. Early work on the extinction law, carried out by Stebbins and Whitford and others in the 1940's, suggested that, in the optical and near infrared at any rate, the extinction A_λ has a λ^{-1} dependence on wavelength. It is this dependence on extinction which suggests that the material responsible consists of small particles having dimensions $a \sim$ an optical wavelength (further confirmation is provided by interstellar polarization; see below). Observations suggest that, in the visible, there is about 1 mag extinction per kiloparsec through the dust layer in the Galaxy. Now

$$\tau_V = \int_{\text{pathlength}} N\sigma_{\text{ext}} dx = \frac{\log_e 10}{2.5} A_V,$$

where N is the number particles responsible for extinction per unit volume and σ_{ext} is their extinction cross-section. Just to get an estimate of N, we suppose that all the particles are identical and that their distribution in space is uniform. We therefore have that

$$N\sigma_{\text{ext}} \sim 3 \times 10^{-20}\,\text{m}^{-1};$$

the agents responsible for interstellar extinction must therefore satisfy this constraint, as well as the Oort limit (1.21) and the observed wavelength-dependence $A_\lambda \propto \lambda^{-1}$ in the optical (see Table 6.1).

We know that there are free electrons in interstellar space. Suppose, then, that interstellar extinction is due to Thomson scattering by free electrons, for which the relevant cross-section is the Thomson cross-section σ_T. This leads to the required electron number density $4.5 \times 10^7\,\text{m}^{-3}$ in interstellar space and a corresponding space density of electrons of $4.1 \times 10^{-22}\,\text{kg}\,\text{m}^{-3}$. However, at

Table 6.1: Possible candidates for cause of interstellar extinction

Ingredient	Extinction law	Cross-section	ρ (kg m^{-3})
Free electrons	λ^0	$\sigma_{\rm T}$	4.1×10^{-22}
Molecules	λ^{-4}	$\sim \pi a^2$	4.4×10^{-26}
Small ($a \ll \lambda$) particles	λ^{-4}	$\ll \pi a^2$	$\gg 1.5 \times 10^{-27}$
Large ($a \gg \lambda$) particles	λ^0	πa^2	$\gg 1.5 \times 10^{-27}$
Particles with $a \sim \lambda$	λ^{-1}	$Q_{\rm ext}\pi a^2$	1.2×10^{-23}

the photon energies of interest in this context, the scattering cross-section is independent of wavelength; hence although the required density of electrons is consistent with the Oort limit, this mechanism for interstellar extinction is ruled out. Furthermore, large ($a \gg \lambda$, where λ here is the wavelength of visible light) particles also display neutral extinction, so these can also be ruled out.

For the same reason, scattering by molecules (e.g. CO) can not be responsible for interstellar reddening and extinction. Again the Oort limit is satisfied, since the requirement is for ~ 1 molecule m^{-3}, corresponding to a mass density of $\sim 4.4 \times 10^{-26}$ kg m^{-3} for CO. However, the extinction law would be that corresponding to Rayleigh scattering, $\propto \lambda^{-4}$. The same wavelength-dependence results from scattering by small ($a \ll \lambda$) dielectric particles [cf. Eq. (3.28)]. Furthermore, from Eq. (3.28), and taking $0.01\,\mu$ silicate grains for which $m \simeq 1.6$ at optical wavelengths, we find $\sigma \simeq 1.7 \times 10^{-20}$ m^2, so that the number density of such grains in the interstellar medium would be ~ 2 m^{-3}. The corresponding average mass density is $\sim 3 \times 10^{-20}$ kg m^{-3}, well *above* the Oort limit.

Particles having $2\pi a/\lambda \simeq 1$ provide the only means of giving rise to a λ^{-1} extinction law and at the same time, satisfying the Oort limit. If, for example, $a \simeq 0.1\,\mu$m, then $\sigma \simeq 3 \times 10^{-14} Q_{\rm ext}$, where $Q_{\rm ext}$ will be of order unity; the corresponding number density of particles is therefore $\sim 10^{-6}$ m^{-3}. If these particles have bulk density 3000 kg m^{-3}, then the averaged-out mass density is $\sim 10^{-23}$ kg m^{-3}, comfortably below the Oort limit. Support for the value of a inferred from extinction and reddening comes from the polarization of starlight, discussed in Section 6.7.

6.4.2 Constraints from the gas-to-dust ratio

If we compare the inferred density of interstellar dust grains, estimated above, with the density of the interstellar gas, we find that the *gas-to-dust ratio*

$$\frac{\rho(\text{dust particles})}{\rho(\text{gas})} \sim 10^{-2}.$$

This ratio implies that a large fraction of heavy elements (i.e. other than hydrogen and helium) must be tied up in dust grains. Most astrophysical objects such as the Sun, HII regions etc. seem to have a fairly standard chemical composition, consisting of 73 per cent hydrogen, 25 per cent helium and 2 per cent of heavier elements by mass. These standard 'cosmic abundances' are listed in Table 6.2 below. Since elemental abundances generally decrease with increasing atomic number (a notable exception to this rule is iron) we are constrained to assemble cosmic dust grains from elements at the lower end of the periodic table. This suggests carbon, nitrogen, oxygen, magnesium, silicon, sulphur and iron, and compounds of these with the others, as these are the only heavy elements present in reasonable abundance ($\gtrsim 10^{-4}$ times the abundance of hydrogen, by number); obviously we are neglecting the inert gases He and Ne in this context. An early candidate for the grain material was water ice but if this were the case, heavily reddened stars should show an absorption feature at 3.1 μm wavelength, characteristic of the stretching of the O–H bond. Such a feature is not in general seen, except in those cases in which the line of sight to a star passes through a molecular cloud (see Section 6.9).

6.4.3 Depletion of elements

We can determine the abundances of the elements in interstellar space because of the interstellar absorption lines in stellar spectra (see Section 2.3). Eq. (6.2) gives the observed flux density S_ν in terms of the optical depth τ_ν. Substituting this in Eq. (2.14) for the equivalent width W gives

$$W_\lambda = \frac{\lambda^2}{c} \int (1 - \exp[-\tau_\nu])\, d\nu. \tag{6.11}$$

This relationship essentially defines the curve of growth, which allows the determination of the abundance of absorbing atoms along the line of sight. If the absorption line is optically thin ($\tau_\nu \ll 1$) we can expand the exponential in Eq. (6.11) so that

$$W_\lambda = \frac{\lambda^2}{c} \int \tau d\nu = \frac{\lambda^2 N \sigma}{c}, \tag{6.12}$$

Table 6.2: Depletion of elements in interstellar space. From W. W. Duley, *Quarterly Journal of the Royal Astronomical Society*, Vol. **25**, 109 (1984)

Species	Solar abundance (H = 1)	ISM abundance (H = 1)	Depletion	Fraction in dust
H	1	–	–	–
He	8.5×10^{-2}	–	–	–
C	3.7×10^{-4}	1.3×10^{-4}	0.35	0.65
N	1.0×10^{-4}	5.0×10^{-5}	0.50	0.50
O	6.8×10^{-4}	5.4×10^{-4}	0.79	0.21
Na	1.7×10^{-6}	2.1×10^{-7}	0.12	0.88
Mg	3.5×10^{-5}	1.0×10^{-6}	0.03	0.97
Si	3.5×10^{-5}	8.2×10^{-7}	0.02	0.98
P	2.7×10^{-7}	1.1×10^{-7}	0.41	0.59
S	1.6×10^{-5}	8.2×10^{-6}	0.51	0.49
Cl	4.4×10^{-7}	9.9×10^{-8}	0.23	0.77
K	1.1×10^{-7}	1.0×10^{-8}	0.09	0.91
Fe	2.5×10^{-5}	2.7×10^{-7}	0.01	0.99

where N is the column density of absorbers and σ is the absorption cross-section of the atom or ion under consideration, integrated over the actual width of the line. Note how, in the limit of small optical depth, the equivalent width is simply proportional to the column density of absorbers; this is referred to as the *linear* portion of the curve of growth. Eq. (6.11) is used to determine the abundance of elements and ions in interstellar space.

Some typical results giving the abundances of selected elements in interstellar space are given in Table 6.2, although it should be borne in mind that there are differences in different lines of sight. However, a glance at Table 6.2 leads us to the conclusion that, in general, the heavy elements are *under-represented*–or *depleted*–in interstellar space compared with typical cosmic abundances. One possibility of course is that the chemical composition of the interstellar gas is somehow intrinsically different from the standard set of abundances. However, it is more usual to conclude that the depletion of elements we see in the interstellar gas is due to the fact that the missing elements are locked up in dust grains, hence the final column in Table 6.2. The other possibility of course is that the depleted elements are tied up in molecules as well as (or even rather than) grains. For example we might expect that a substantial amount of the missing carbon may be in the form of CO, or other carbon-bearing molecules (such as polycyclic aromatic hydrocarbons–PAHs; see Chapter 8). The information provided by the depletion of elements in the interstellar gas seems therefore not inconsistent with that provided by the gas-to-dust ratio.

6.4.4 Dust-related absorption-features

We have already seen (see Figs. 6.1, 6.2) that there is a broad absorption feature in the ultraviolet spectra of heavily reddened stars, and there is no doubt that successful identification of this feature would be a significant step forward in the understanding of interstellar dust. The obvious identification is small ($\lesssim 0.05\,\mu$m) spherical graphite particles, which do indeed have a strong feature at this wavelength (see Fig. 3.10). However, there are problems with the obvious candidate, such as the fact that graphite is unlikely to grow as spheres and that the size of the particles has to be constrained: note that the central wavelength and profile of the absorption feature changes as we go to larger grains. An alternative identification is OH^- ions on the surface of small silicate grains. Both identifications predict the presence of other, weaker absorption features at infrared wavelengths, but neither identification can be ruled out at present.

More secure is the identification of an absorption feature at $9.7\,\mu$m, which is seen in directions in which the reddening is extremely heavy. One such line of sight is the Galactic centre, for which the visual extinction is extremely high, amounting to $A_V \simeq 30$. Fig. 6.3 shows the spectrum of the infrared source close to the centre of the Galaxy. There are several absorption features, the most prominent of which is that centred on $9.7\,\mu$m and attributed to silicates (see Fig. 3.9); note also the presence of the other, weaker, silicate feature at $18\,\mu$m. The weaker features at 3 and $3.4\,\mu$m may be related to OH stretch in hydrated silicates and to organics respectively. We note that there is no absorption feature centred on $11.5\,\mu$m, suggesting that silicon carbide is not a major constituent of interstellar dust despite the fact that it is commonly seen in circumstellar environments (see Chapter 7).

Although water ice was one of the first identifications proposed for interstellar dust, it turns out that ice is not present to any great extent in interstellar space, either as a prime constituent or on the surfaces of interstellar dust grains. However both water ice (absorption feature at $3.07\,\mu$m) and solid carbon monoxide ($4.8\,\mu$m) are seen in absorption in molecular clouds.

In the optical band there are a number (over 100) of absorption features which, although first detected in the 1920's, have hitherto defied identification; these features are commonly referred to as 'diffuse interstellar bands'. The central wavelengths of these features range from the blue (e.g. the fea-

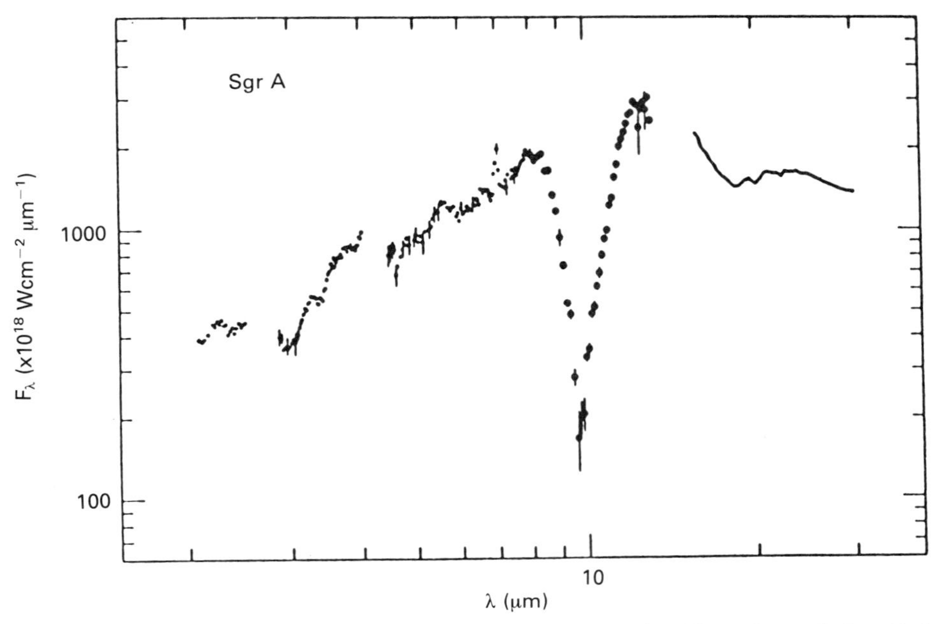

Figure 6.3: Interstellar absorption features in the direction of the Galactic centre. After P. F. Roche, *Dust in the Universe*, p 415, Eds M. E. Bailey & D. A. Williams, Cambridge University Press (1988).

ture at 4428 Å) to the far red (e.g. 8650 Å) and there is some evidence that there are distinct 'sets' of features (i.e. they do not all have the same carrier). The equivalent widths of these features generally correlate well with interstellar reddening [as measured, for example, by $E(B-V)$], although there are exceptions to this rule in that some stars are heavily reddened but display no diffuse bands. This correlation suggests either that the grains responsible for interstellar extinction are also responsible for the diffuse bands, or that the material responsible for the diffuse bands is well-mixed with the grains responsible for reddening. It seems significant that there is no evidence for diffuse bands in *circumstellar* dust: apparently they are exclusively an interstellar phenomenon. A selection of optical spectra displaying diffuse bands is shown in Fig. 2.5.

6.4.5 The composition and size distribution of interstellar dust

If we take an 'average' interstellar extinction curve (bearing in mind that such a concept may not be entirely meaningful) it is found that it can be well fitted by a mixture of two grain populations, one consisting of silicate grains, the other of carbon grains; in this case the 2175 Å extinction feature is attributed to carbon, despite the reservations made above. In each case the grain size distribution, the number of grains per unit volume having radius in the range $a \rightarrow a + da$, is of the form

$$n(a)da \propto a^{-3.5}da,$$

which is close to that which is characteristic of grain-grain collisions [see Eq. (5.37) and Appendix B]. The amount of carbon, silicon etc. tied up in the resultant interstellar grain population is also consistent with the depletion of these elements in the interstellar gas (see Table 6.2). This model of the interstellar grain population is frequently referred to as the MRN model, after the initials of the workers who devised it. In general we do not expect the necessary grain-grain collisions to occur in interstellar space, where the number of grains per unit volume is far too low ($\simeq 10^{-6}\,\mathrm{m}^{-3}$; see Section 6.4.1). The implication therefore is that interstellar grains form in the environments of stars where the grain size distribution is established, and are then ejected into interstellar space. This, as we shall see, is also consistent with other requirements for the origin of interstellar dust.

An alternative view is that there is no separate population of silicate and carbon grains, but that silicate grains are coated with a thin mantle of amorphous carbon. Variations in the interstellar extinction law in various directions are, in this case, attributed to variations in the thickness of the carbon mantle, while the 2175Å feature is attributed to absorption by small silicate grains (see above).

In summary, there is no serious doubt that silicates and carbon are major constituents of the interstellar dust population, although whether these materials form distinctly separate populations or are intimately connected as core-mantle grains remains to be seen. As always, the understanding of interstellar grains is dependent on laboratory measurements of the optical properties of relevant materials, and on an understanding of the physics of small particles and of processes on grain surfaces.

6.5 Scattering by interstellar dust

6.5.1 Reflection nebulae

There are many instances in which a star illuminates a cloud of nearby (to the star) interstellar dust grains. Such a phenomenon is referred to as an interstellar reflection nebula and it has long been recognized that the

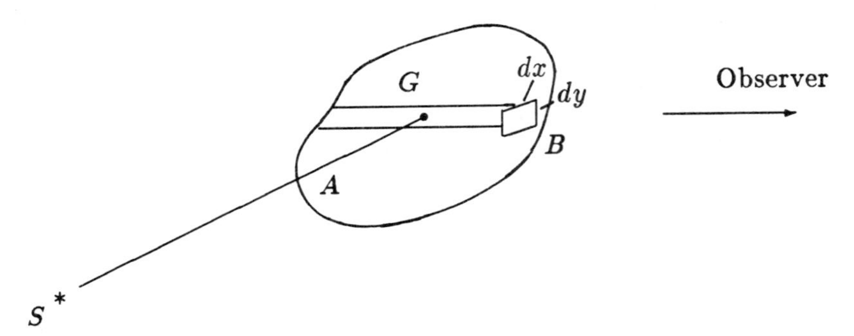

Figure 6.4: Geometry for deriving Hubble's relation.

observed angular dimensions of such nebulae are related in a simple way to the apparent magnitude of the illuminating star. This relationship is referred to as Hubble's relation (not to be confused with Hubble's law, to be discussed in Chapter 9).

That a relationship like Hubble's is to be expected may be seen from the following simple argument. Suppose we have a star S, situated near a small cloud of interstellar dust in which the grains, assumed identical, are distributed uniformly (see Fig. 6.4). Light from the star enters the nebula at A *en route* to grains at G in the interior of the cloud, whence it is scattered in the direction GB towards the observer; the scattering angle is θ and we assume for simplicity that the scattering is isotropic. The star is close to the cloud, but not so close that the values of the distance SG ($= r$) and of the angle θ differ for different locations in the cloud. The elements of length dx and dy define elements of length in the plane of the sky.

If the absolute luminosity of the star is L, the flux reaching G is

$$f_{\mathrm{G}} = \frac{L}{4\pi r^2} e^{-\tau_{\mathrm{AG}}}, \tag{6.13}$$

where τ_{AG} is the extinction optical depth along the path AG. At G light is scattered in the direction of the observer. From Eq. (2.2) the amount of light removed by scattering is

$$f_{\mathrm{G}} n_{\mathrm{d}} \pi a^2 Q_{\mathrm{sca}} ds = f_{\mathrm{G}} d\tau_{\mathrm{sca}},$$

where $ds = dx \csc\theta$ is an element of length along SAG and $d\tau_{\mathrm{sca}}$ is the optical depth due to scattering only. We can express τ_{sca} in terms of the extinction optical depth τ by means of the albedo ϖ, defined in Eq. (3.19). Thus the scattered flux from G, emerging at B, is

$$= \frac{f_{\mathrm{G}}}{4\pi} \varpi d\tau \tag{6.14}$$

$$= \frac{L}{4\pi r^2} e^{-\tau_{AG}} e^{-\tau} \frac{\varpi}{4\pi} d\tau. \tag{6.15}$$

The observed flux emerging from B is then given by integrating over the depth of the cloud, along CB:

$$df = \frac{L}{4\pi r^2} \frac{\varpi e^{-\tau_{AG}}}{4\pi D^2} dx dy \int_{\text{pathlength}} e^{-\tau} d\tau \tag{6.16}$$

$$= \frac{L}{4\pi r^2} \frac{\varpi e^{-\tau_{AG}}}{4\pi D^2} dx dy [1 - e^{-\tau}]. \tag{6.17}$$

We can express $L/4\pi D^2$ in terms of apparent luminosity S of the star [see Eq. (1.15)]; also $dxdy/D^2 = d\Omega$, the element of solid angle at B. Thus the *intensity* of the nebula at B (e.g. in W m^{-2} Hz^{-1} sr^{-1}) is given by

$$I = df/d\Omega \tag{6.18}$$

$$= S e^{-\tau_{AG}} [1 - e^{-\tau}] \frac{\varpi D^2}{4\pi r^2}. \tag{6.19}$$

From Fig. 6.4 we see that

$$\alpha = \frac{r \sin\theta}{D}, \tag{6.20}$$

where α is the observed angular separation between S and B. Substituting in Eq. (6.18) we get

$$I = S e^{-\tau_{AG}} [1 - e^{-\tau}] \frac{\varpi \sin^2\theta}{4\pi \alpha^2}. \tag{6.21}$$

Using Eq. (1.9), we can express the intensity of the nebula and the apparent luminosity of the star in magnitudes:

$$m_{\rm n} = m_* + 5\log\alpha - 2.5\log[\sin^2\theta] - \left\{ 2.5\log\frac{[1 - e^{-\tau}]\varpi}{4\pi} + \frac{2.5\tau_{AG}}{\log_e 10} \right\}, \tag{6.22}$$

where $m_{\rm n}$ is usually expressed in magnitudes per square arcsecond. The quantity in curly brackets in Eq. (6.22) depends on the properties of the grains in the cloud and their distribution, while the $\sin^2\theta$ term depends on the geometry. Thus Eq. (6.22) applies to a specific star-nebula combination. For a large sample of combinations we will obviously have different values of $\sin^2\theta$ etc. and we therefore take average values of these quantities; note that τ enters only through the factor $[1 - \exp(-\tau)]$, which does not differ very much from unity, and that $\langle \sin^2\theta \rangle = 2/3$. Thus

$$m_{\rm n} \simeq m_* + 5\log\alpha + \text{constant}. \tag{6.23}$$

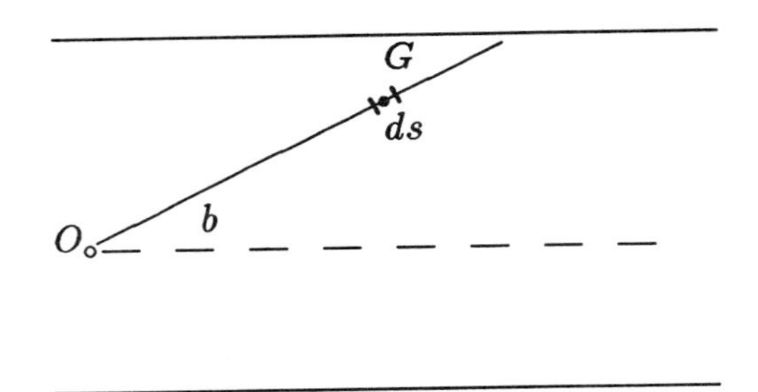

Figure 6.5: Geometry for scattering by interstellar dust.

Clearly the intensity of the scattered light falls away rapidly the further we go from the star. Eventually the intensity declines to a value that is less than the background light, at which point the nebula is no longer visible. The extent α_1 of the reflection nebula is therefore determined by the level of the background which, for a given method of detection (e.g. a photographic plate) is a constant. Thus we arrive at Hubble's relation

$$m_* = 5 \log \alpha_1 + \text{constant}. \tag{6.24}$$

Nebulae that obey Eq. (6.24) may be said to be *radiation bounded*, since the visible edge of the nebula is determined by the intensity of the radiation it emits; other reflection nebulae may be *matter bounded*, in which the visible edge of the nebula is determined by an abrupt decline in the grain population. We would not expect matter bounded reflection nebulae to follow the Hubble relation.

6.5.2 Scattering of background starlight by interstellar dust

In addition to the scattering of starlight by a specific dust cloud, we also get scattering of background starlight by the general population of interstellar grains. Again we shall assume isotropic scattering and identical grains for simplicity, and that the grain distribution is uniform, with n_d grains per unit volume. We consider a line of sight from the observer O through the interstellar dust layer, and making an angle b with the plane of the Galaxy; i.e. we are looking in a direction having Galactic latitude b (see Fig. 6.5).

The background light consists of two components, namely the background starlight and the scattered light; let the intensity of this background be I. We consider an element of length ds at a point G along the line of sight, along which the extinction optical depth is $d\tau$. The intensity as a result of scattering by grains at G is given, as in the previous section, by

$$dI = \frac{I}{4\pi} \varpi d\tau, \tag{6.25}$$

where ϖ is again the albedo and isotropic scattering has been assumed. Only a fraction $e^{-\tau}$ of this reaches the observer, so the total (scattered) intensity measured at O is

$$I' = \frac{I\varpi}{4\pi}\int_0^{\tau_0} e^{-\tau}d\tau$$

where τ_0 is the total optical depth along the line of sight. Note the difference between I and I': the latter is the intensity of scattered light only, whereas the former includes both scattered light and stellar background; obviously $I > I'$. Carrying out the integration we get:

$$I' = \frac{I\varpi}{4\pi}[1 - e^{-\tau_0}]. \tag{6.26}$$

Close to the Galactic plane, $\tau_0 \gg 1$ and so the term in square brackets in Eq. (6.26) is very close to unity. Thus

$$\frac{I'}{I} \simeq \frac{\varpi}{4\pi}. \tag{6.27}$$

In principle both I and I' can be measured separately, leading to a value for the albedo ϖ.

More rigorously we should take account of the scattering phase function for the grains $S(\theta)$. In this case Eq. (6.25) becomes

$$dI = \frac{\varpi}{4\pi}d\tau \int_0^{2\pi} I(\alpha)S(\alpha)d\alpha, \tag{6.28}$$

and

$$I' = \varpi[1 - e^{-\tau_0}]\int_0^{2\pi} I(\alpha)S(\alpha)d\alpha \tag{6.29}$$

$$= \varpi[1 - e^{-\tau_0}]I\int_0^{2\pi} S(\alpha)d\alpha \tag{6.30}$$

if $I(\alpha)$ is independent of α. Thus in this case

$$\frac{I'}{I} \simeq \varpi\int_0^{2\pi} S(\alpha)d\alpha.$$

6.5.3 Scattering when the illuminating source is variable

It sometimes happens that the radiation scattered by dust grains originates in a variable source, so that the scattered light is itself variable. Such behaviour is further complicated by the fact that the spectral shape or effective temperature of the illuminating object may be variable as well. By analogy

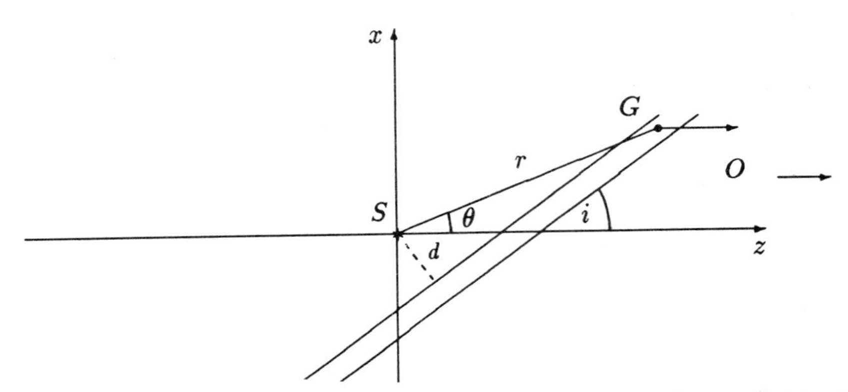

Figure 6.6: Scattering of light from a variable source by a sheet of interstellar dust.

with the corresponding acoustic case this phenomenon is often referred to as a 'light echo'.

In order to illustrate the kind of effect we are likely to get, we consider the situation shown in Fig. 6.6, in which a thin plane sheet of interstellar dust, inclined at angle i to the line of sight, lies between a variable star S and a distant observer O; the perpendicular distance from the star to the dust sheet is d. For the stellar light curve we take the simplest possible case, namely a star that has zero luminosity initially, then at a given time instantaneously 'switches on' so that the resultant luminosity is L_0. The star is assumed to radiate isotropically and we shall not concern ourselves with the spectral distribution of the radiation. In order to set the timescale we suppose that the observer defines the instant she sees the star switching on as $t = 0$. Thus

$$
\begin{aligned}
L &= 0 & (t < 0) \\
L &= L_0 & (t > 0).
\end{aligned}
$$

Because light scattered off the dust sheet has further to travel from the star, via the dust, to the observer, there will be a delay between the observer seeing the star switching on at $t = 0$, and seeing the scattered light reflected off the dust sheet. In other words light takes longer to travel the path star–grain–observer than to travel directly to the observer along SO.

What does the observer see at time $t > 0$? To see this we concentrate on one specific ray, SG in Fig. 6.6, which leaves that star at angle θ to the line of sight and travels a distance r to the dust sheet before being scattered towards the observer. If the observer is so far away that light from both S and G travel towards the observer in parallel lines, the path taken by the scattered ray is $r(1 - \cos\theta)$ longer than the direct line of sight SO to the observer. From the observer's point of view, therefore, there is a time delay, given by

$$\delta t = r(1 - \cos\theta)/c, \tag{6.31}$$

between the sudden appearance of the star at $t = 0$ and the appearance of scattered light from a grain at G. Note that, because of symmetry about the line of sight, this delay applies to all grains such that the angle GSO $= \theta$. Indeed, the scattered light arriving at the observer at any given time t will have arisen from the surface of a paraboloid of revolution defined by

$$r = \frac{ct}{1 - \cos\theta}. \tag{6.32}$$

In order to determine the appearance of the scattered light from the observer's point of view it will be more convenient to use cartesian co-ordinates, with the z-axis towards the observer and the x and y axes in the plane of the sky (see Fig. 6.6); the star is at the origin of co-ordinates. We therefore need the equation of a paraboloid of revolution corresponding to Eq. (6.32); in cartesian co-ordinates we have

$$x^2 + y^2 = c^2t^2 + 2zct. \tag{6.33}$$

In the xz plane the equation of the dust sheet is just

$$x = z\tan i - d\sec i. \tag{6.34}$$

What the observer sees is bounded by the intersection of the paraboloid and the dust sheet, projected on to the plane of the sky (the $x-y$ plane). We can determine this projection by eliminating z between Eqs. (6.33) and (6.34) to get

$$(x - ct\cot i)^2 + y^2 = c^2t^2\csc^2 i + 2ctd\csc i. \tag{6.35}$$

Note that this is just the equation of a circle, whose centre is displaced from the position of the light source by $ct\cot i$. The observer sees a circle of scattered light, the radius

$$\rho = ct\left[\csc^2 i + \frac{2d\csc i}{ct}\right]^{1/2} \tag{6.36}$$

of which expands in size and which recedes from the source of illumination. For the simple case in which $d = 0$, the radius of the circle of scattered light is $\rho = ct\csc i$; note that the expansion *rate* is $\dot{\rho} = c\csc i$ which can be $\gg c$. Such *superluminal* effects are common for the sort of situation we have described here.

The effects discussed here have been seen in a number of situations, which we now briefly describe.

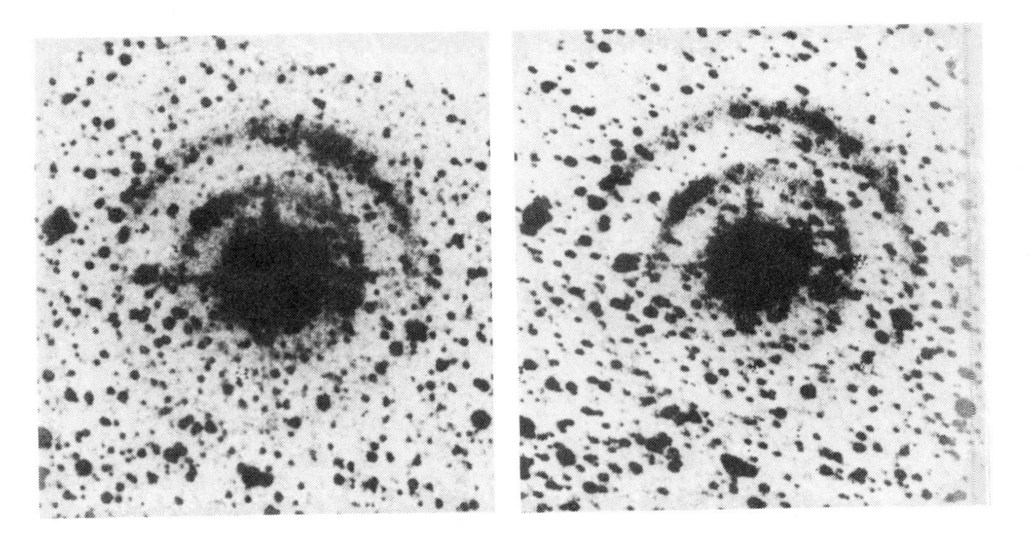

Figure 6.7: Light echoes around supernova SN1987A. After C. Gouiffes et al., *Astronomy & Astrophysics*, Vol. **198**, L9 (1988).

Supernova 1987A

The eruption of the supernova 1987A in 1987 February was without doubt one of the most significant astronomical events of the 20 th century. SN 1987A was the first nearby supernova for over 300 years and the first to be studied in great detail with the enormous battery of sophisticated detectors available to the late-twentieth century astronomer. For example, the copious burst of neutrinos emitted during the second or so it took for the core of the progenitor star to collapse were detected by at least two neutrino 'observatories'–the first detection of neutrinos from a celestial object other than the Sun.

In the context of light scattering by dust particles, however, the crucial observations were carried out over a year after the supernova erupted. In February 1988 two rings of light, centred roughly on the supernova, were observed (see Fig. 6.7). Further observations revealed that these rings were expanding at a rate of about 3 seconds of arc per month. Optical spectroscopy of the rings showed that, spectroscopically, they resembled the supernova itself, as it had been at *maximum light* a year previously. The rings were clearly an 'echo' of the 1987 eruption, reflected off sheets of interstellar dust in the Large Magellanic Cloud (see Fig. 6.8). Note that we see rings, rather than filled circles, of reflected light; the light signal from SN1987A was not maintained after the initial outburst and the supernova went into decline.

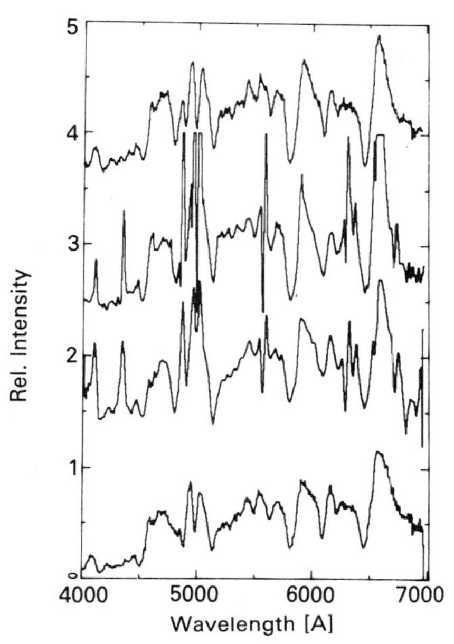

Figure 6.8: Top and bottom traces are the optical spectra of SN1987A close to maximum light; the two central spectra are of the inner ring (upper centre) and outer ring (lower centre) of the light echo observed around the supernova in March 1988; see Fig. 6.7. After C. Gouiffes et al., *Astronomy & Astrophysics*, Vol. **198**, L9 (1988).

From the observed properties of the rings (rate of expansion etc.), and the geometry described by Eq. (6.35), the values of i and d may be determined, while the observed brightness of the rings is determined by the grain content. Note also that, at the distance of the supernova (56 kpc), the observed angular expansion rate corresponds to an expansion *speed* of about $30\,c$.

Nova GK Persei 1901

A similar phenomenon had been seen decades earlier, shortly after the eruption of the nova GK Persei in 1901. This nova erupted in February 1901 and, some months after outburst, expanding nebulosities, in the form of bright 'arcs' about the nova, were detected photographically. Since the distance of the nova had been reliably determined to be 470 pc so the velocities at which the nebulosity expanded could be determined directly. Although not as well observed as the light echo around SN1987A, the distance of the scattering sheet was estimated to be some 14 pc from the nova.

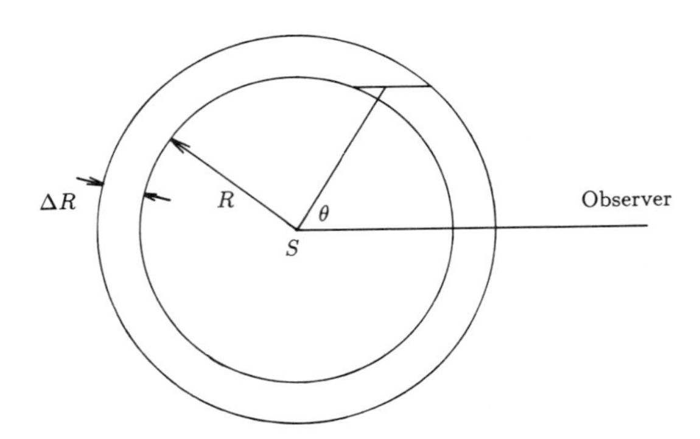

Figure 6.9: Scattering of light by dust shells around RS Puppis.

RS Puppis

Cepheid variable stars are pulsationally variable stars the luminosities of which vary with precisely defined periods in the range 10–100 days. Their importance lies in the fact that the luminosities of Cepheid variables depend in a simple way on the light curve period: the determination of the light curve period from observation determines the absolute luminosity and hence the distance–provided that the period-luminosity law has been calibrated. The calibration can be performed by observing Cepheid variables in clusters of stars of known distance, although independent checks are obviously always desirable.

RS Puppis is a Cepheid variable star which has a pulsational period of 41.4 days. It has a unique place amongst the Cepheids in that it possesses a number of concentric 'rings' of dust which scatter light from the central star. As the star varies with its characteristic light curve, the scattered light from the rings of dust also varies, but the variation is out of phase with the variation of the star itself. Crudely speaking, if a dust shell has radius R, the variation of the light scattered from the dust will, because of the finite velocity of light, be delayed relative to the variation of the star itself by $\sim R/c$. The observational determination of this phase difference determines R which, with the measured angular diameter of the dust shell, gives the distance of RS Puppis. However, this neglects the effect of the scattering phase function $S(\theta)$, the effect of which is to make the shell appear smaller than it really is. To see why this is so, note that, for a spherical dust shell of radius R and thickness $\Delta R(\ll R)$, centred on the star at S, the observed intensity of the scattered light $I \propto \sec\theta S(\theta)$ (see Fig. 6.9). If we suppose that the scattering properties of the dust are described by the Heyney-Greenstein scattering function [Eq. (3.27)], we can see by differentiation that the maxi-

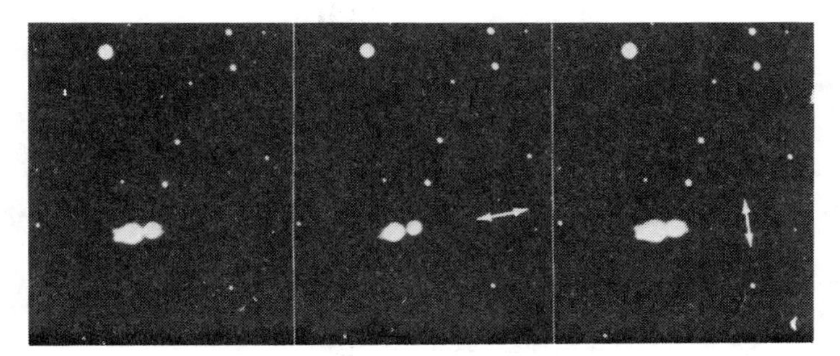

Figure 6.10: A bipolar reflection nebula, the 'Egg nebula'. The spectral coverage is 600–700 nm. Left, no polarizer. Centre and right, images in polarized light with electric vector oriented as shown. Photographed by Richard H. Cromwell with the 2.2 m telescope of the University of Arizona. Steward Observatory Photograph.

mum intensity occurs when

$$\cos\theta_{\rm m} = \frac{1+g^2}{5g},$$

where g $(= \langle\cos\theta\rangle)$ is the average of $\cos\theta$ averaged over the scattering phase function. The *observed* size of the shell $R_{\rm obs} \simeq R\sin\theta_{\rm m}$ therefore depends on g but for forward-scattering grains having $g \simeq 0.7$, $R_{\rm obs} \simeq 0.9R$. Thus while the variability of the dust shells of RS Puppis offer a novel means of determining the distance of a Cepheid variable there is uncertainty in the method following from the uncertainty in g. Such uncertainty would be removed if we could measure the thermal infrared emission from the dust, because for spherical grains the infrared emission is isotropic.

6.5.4 Compact reflection nebulae

We now consider the scattering of starlight by dust grains that are intimately associated with the illuminating star. It is unusual for reflection nebulae associated with specific stars to possess spherical symmetry: often the nebulosity is the result of a *bipolar outflow*, in which material is ejected by the star in two well-collimated directions. In some cases the outflow may be directly seen as a result of light scattered from dust grains therein and, invariably, the outflow is also detectable by virtue of strong molecular emission (especially CO). A bipolar outflow is shown in Fig. 6.10.

In order to illustrate the essential features of the reflected light from a bipolar outflow we consider the emission of grain-forming material in the form of two narrow co-axial cones, the apices of which meet at the position

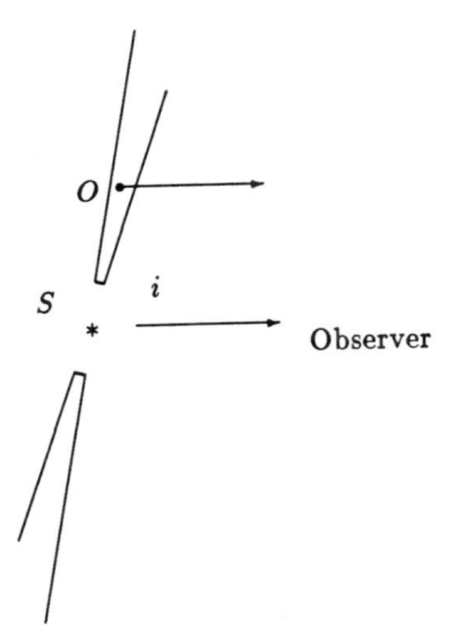

Figure 6.11: Geometry for calculating the light scattered from a bipolar outflow.

of the star (see Fig. 6.11); the cones are inclined at angle i to the plane of the sky. We assume that the cones are thin enough that we can neglect the extinction of scattered light, although the light from the illuminating star is extinguished along the length of the cone. The grains will condense at some distance $r_{\rm c}$ from the star, where $r_{\rm c}$ will depend on the grain material and on the properties of the star itself (see Section 7.5.2 below). In an outflow the number density of grains $n_{\rm d}$ will vary as $n_{\rm d} \propto r^{-2}$ [see Eq. (7.11) below] so the optical depth along a line of sight from the star to a point O distance r from the star (see Fig. 6.11) is

$$\tau_{\rm O} = \int_{r_{\rm c}}^{r} n_{\rm d}(r) Q_{\rm ext} \pi a^2 dr.$$

The apparent luminosity of the star as seen by a grain at O is

$$\frac{L_*}{4\pi r^2} \exp[-\tau_{\rm O}].$$

The scattered intensity at angular separation $\alpha = b/D$, where D is the distance of the system and $b = r \sin i$ is the projected separation of O from the star, is

$$I(\alpha) \simeq \frac{L_*}{4\pi r^2} \exp[-\tau_{\rm O}] n_{\rm d}(r) S(\pi/2 - i). \tag{6.37}$$

We note that (i) the intensity of the scattered light will fall substantially (because of the $\exp[-\tau_O]$ term) as we go further from the star and (ii) since the grains are likely to be forward-scattering the component of the outflow leaning towards the observer is brighter than the component leaning away from the observer. Such differences in the brightnesses of the components of bipolar outflows are commonly seen (see Fig. 6.10).

We also recall here the polarization properties of the scattered light (Section 3.4.3). The star, scattering dust and observer in the case of the reflection nebula in Fig. 6.10 define the scattering plane and, as already discussed, the scattered light is expected to be linearly polarized at right angles to this plane. As is evident from Fig. 6.10, this is borne out by polarimetric observations of the nebula: note that the image (centre) for which the electric vector is perpendicular to the scattering plane is much more prominent than that (right) in which the electric vector is in the plane.

6.5.5 X-ray scattering

The possibility that interstellar dust grains might scatter X-radiation was first mooted in 1965, at about the time that serious astronomical observations of X-ray sources were being undertaken for the first time. However, observational verification of the early predictions had to await the availability of satellite X-ray observatories that had an imaging capability, such as the *Einstein* and *EXOSAT* observatories. This is because the scattering of X-rays gives rise (as in the optical case discussed in Section 6.5.1) to extended emission–a halo–which has a specific spatial distribution, and simply measuring a flux (as might have been possible using earlier facilities such as *Uhuru*) is not sufficient to verify the presence of X-ray scattering.

A crucial parameter in discussing the scattering of X-rays by interstellar dust is the differential scattering cross-section, which determines the amount of radiation scattered into a solid angle $d\Omega$ around the incident direction. There is obviously a close similarity between the differential scattering cross-section and the scattering phase function $S(\theta)$ of Chapter 4 (cf. Fig. 3.6). As interstellar grains have dimensions typically $\sim 0.1 - 0.2\mu$m (see Sections 6.4 and 6.7), and X-rays have wavelengths of $\leq$ 100 Å, we are in a regime where the parameter $x = 2\pi a/\lambda$ is $\gg 1$. However, this is not to say that we can use the results from Chapter 4 for this limiting case because the details of the scattering process are not the same.

To a good approximation, the real and imaginary parts of the refractive index m at X-ray wavelengths may be written as

$$n' = \Re(m) = 1 - \frac{r_e \lambda^2 n_e}{2\pi}$$

$$n'' = \Im(m) = \frac{\mu\lambda}{4\pi},$$

in which r_e is the classical electron radius, n_e is the number density of electrons in the grain material and μ is the absorption coefficient for the grain. For material of density ρ and molecular weight A, and whose constituent atoms have atomic number Z,

$$n_e = \frac{\rho Z}{m_p A}.$$

For example, olivine ($MgSiO_3$) has $A = 100, Z = 50$; we therefore find that, at X-ray wavelengths

$$\begin{aligned} |n' - 1| &\ll 1 \\ n'' &\ll 1 \end{aligned}$$

for olivine.

In general, when the conditions

$$\begin{aligned} |m - 1| &\ll 1 \\ \frac{2\pi a}{\lambda} &\gg 1 \\ \frac{4\pi a}{\lambda}|m - 1| &\gg 1 \end{aligned}$$

are satisfied the scattering process is referred to as 'anomalous diffraction'. In this case the emerging ray is the combined effect of diffraction and transmission (with negligible absorption) of the incident ray through the particle. Under these circumstances the differential scattering cross-section is

$$\frac{d\sigma}{d\Omega} = 2a^2 x^4 |m - 1|^2 \left(\frac{J_1(\xi)}{\xi}\right)^2 (1 + \cos^2\theta). \quad (6.38)$$

In Eq. (6.38) θ is the scattering angle, $\xi = 2x\sin(\theta/2)$ and $J_1(\xi)$ is the spherical Bessel function of first order; as in Chapter 4, $x = 2\pi a/\lambda$. The total scattering cross-section is obtained by integrating $d\sigma/d\Omega$ over all directions:

$$\sigma_{sca} = 2\pi a^2 x^2 |m - 1|^2; \quad (6.39)$$

the corresponding extinction cross-section is

$$\sigma_{ext} = \frac{8\pi a^2}{3} x \Im(m - 1). \quad (6.40)$$

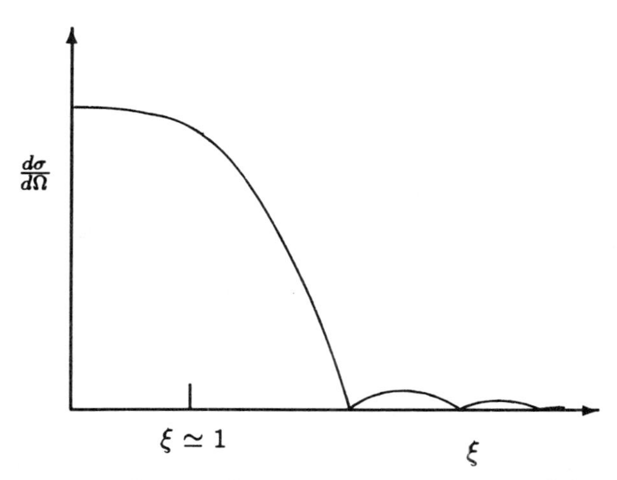

Figure 6.12: The differential scattering cross-section for anomalous diffraction scattering.

As before, the symbol $\Im$ denotes that the imaginary part of the quantity is to be taken.

The dependence of $d\sigma/d\Omega$ in Eq. (6.38) on ξ is shown schematically in Fig. 6.12. For $\xi \lesssim 1$ the dependence of $d\sigma/d\Omega$ on ξ is flat whereas for $\xi \gtrsim 1$ $d\sigma/d\Omega \propto \xi^{-4}$. This means that the scattered intensity falls very rapidly with angle and is negligible for all but the smallest angles. Indeed we can use the dependence of $d\sigma/d\Omega$ on ξ to estimate the size of a scattered halo around a cosmic X-ray source (see Fig. 6.13). Grains along the direct line of sight to the X-ray source scatter radiation out of the line of sight, but grains that are off the line of sight scatter radiation towards the observer, giving rise to the observed halo. If the scattering angle is such that $\xi \gtrsim 1$ then little scattered radiation is observed because the behaviour of $d\sigma/d\Omega$ (see Fig. 6.12) ensures that little or no radiation is scattered at these angles. The observed extent of the halo is thus determined by the condition $\xi \simeq 1$, i.e.

$$\xi = \frac{4\pi a}{\lambda} \sin\left(\frac{\theta}{2}\right) \simeq 1.$$

Since $2\pi a/\lambda \gg 1$ this implies that $\sin\theta/2 \ll 1$ so we can approximate the sine term by $\theta/2$; thus the observed extent θ_1 of the scattered halo is given by

$$\frac{4\pi a}{\lambda}\left(\frac{\theta_1}{2}\right) \simeq 1$$

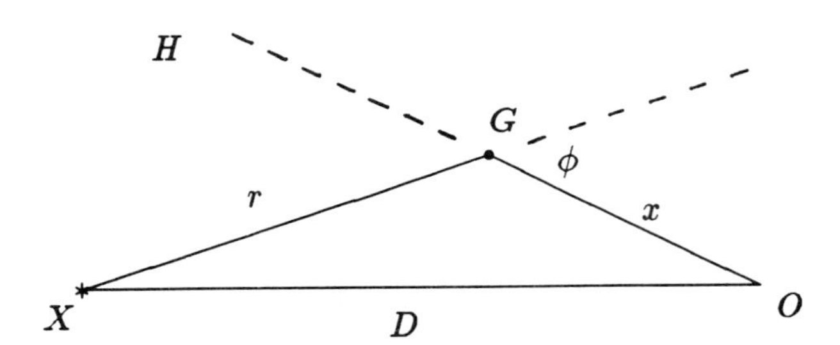

Figure 6.13: Geometry for X-ray scattering by interstellar dust.

or

$$\theta_1 \simeq 5.5 \left(\frac{\lambda}{10\text{Å}}\right) \left(\frac{a}{0.1\,\mu\text{m}}\right)^{-1} \text{arcmin.}$$

We see that X-ray haloes can be expected to be a few arcminutes in extent, with larger haloes for softer X-rays and with smaller scattering grains.

We can also estimate how much X-ray flux to expect in a scattered halo. Suppose we have a point source whose X-ray flux in the absence of interstellar grains would be S. The observed flux will be attenuated by two factors: absorption and scattering by grains, and absorption by the interstellar gas. The dust attenuates the X-rays by an amount $e^{-(\tau_s+\tau_a)}$, where τ_s and τ_a are the optical depths for scattering and absorption by the dust respectively; similarly the attenuation due to the gas is $e^{-\tau_g}$. The observed X-ray flux is thus

$$S = S_0 \exp -[\tau_a + \tau_s + \tau_g].$$

The direct and scattered paths are not identical (see Fig. 6.13), but since the scattering angles are so small it is a good approximation to suppose that the absorption optical depths (for the gas and dust) are the same for the direct and scattered rays. The scattered flux is $S_0(1 - e^{-\tau_s})$ [cf. Eq. (7.5) in Chapter 7] but this is again attenuated because of absorption by the gas and dust. The observed flux in the X-ray halo is thus

$$S_{\text{halo}} = S_0[1 - e^{-\tau_s}]e^{-(\tau_a+\tau_g)}.$$

We can therefore see that

$$\frac{\text{Observed flux in halo}}{\text{Observed source flux}} = e^{\tau_s} - 1 \simeq \tau_s, \qquad (6.41)$$

where the last approximation holds if the scattering optical depth is small.

Calculation of scattered X-ray intensity

The intensity of the scattered X-rays at angular distance α from a point source is calculated as follows (see Fig. 6.13). Suppose we have a source X of X-rays distance D from the observer O. X-rays from the source propagate along the path XG and are scattered by grains at G, through an angle ϕ, in the direction of the observer at O; GO $= x$. If the source has X-ray luminosity L_{X} and the distance XG $= r$, the X-ray flux at G is

$$f_{\mathrm{G}} = \frac{L_{\mathrm{X}}}{4\pi r^2} e^{-\tau_{\mathrm{XG}}} \tag{6.42}$$

where τ_{XG} is the optical depth for the extinction of X-rays along the path XG; as we have already seen this will occur by scattering, and by absorption by both interstellar gas and grains. This is yet another application of the transfer equation Eq. (2.11); the background term is again zero and the source function

$$\mathcal{S} = f_{\mathrm{G}} n_{\mathrm{d}} \frac{d\sigma}{d\Omega} = \frac{L_{\mathrm{X}} n_{\mathrm{d}} e^{-\tau_{\mathrm{XG}}}}{4\pi r^2} \frac{d\sigma}{d\Omega},$$

where n_{d} is the number of interstellar dust grains per unit volume at G. To get the X-ray intensity at angular separation α from the source we integrate along the line OGH:

$$I(\alpha) = \int_{\mathrm{OGH}} \mathcal{S} e^{-\tau_{\mathrm{GO}}} \, d\tau, \tag{6.43}$$

where τ_{GO} is the optical depth for scattering and absorption along the path from G to the observer. Inserting $\mathcal{S}$ in Eq. (6.43) we get

$$I(\alpha) = \int_{\text{path length}} \frac{L_{\mathrm{X}} n_{\mathrm{d}}}{4\pi r^2} \frac{d\sigma}{d\Omega} e^{-[\tau_{\mathrm{XG}} + \tau_{\mathrm{GO}}]} dx, \tag{6.44}$$

where dx is an element of length along GO. Eq. (6.44) can be simplified considerably. We note that

$$\frac{r}{\sin\alpha} = \frac{D}{\sin(\pi - \phi)} = \frac{D}{\sin\phi} = \frac{x}{\sin\theta},$$

and that $\phi = \alpha + \theta$ from geometry; thus

$$\frac{dx}{D} = \frac{\sin\alpha}{\sin^2\phi} d\phi. \tag{6.45}$$

The *observed* X-ray flux from the source is just

$$f_{\mathrm{X}} = \frac{L_{\mathrm{X}}}{4\pi D^2} e^{-\tau_{\mathrm{XO}}}$$

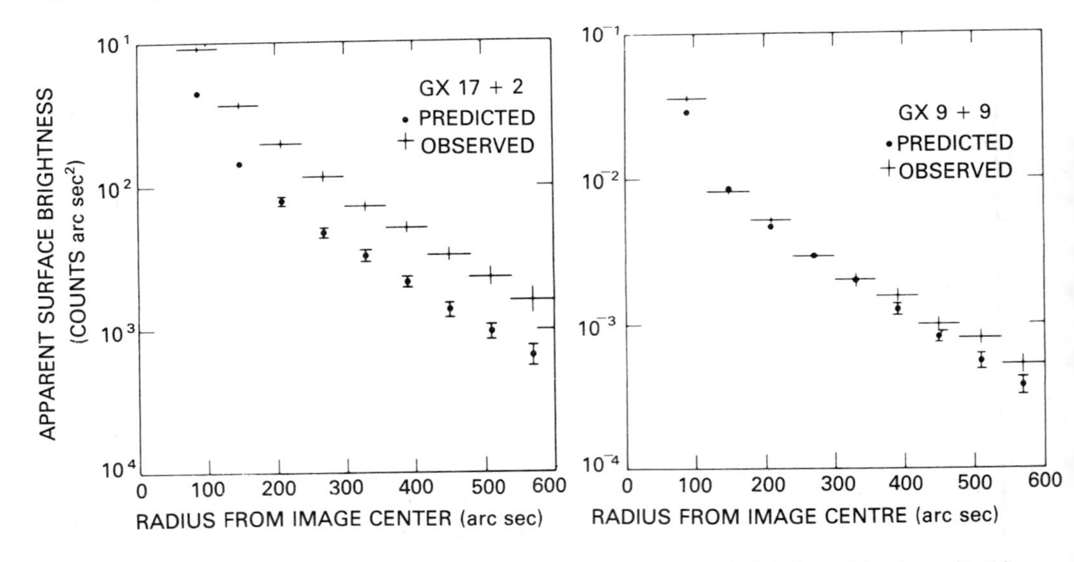

Figure 6.14: X-ray halo around an X-ray source with high reddening (left) and an X-ray source with low reddening (right). After R. C. Catura, *Astrophysical Journal*, Vol. **275**, 645 (1983).

and since the scattering angles are small, we can write $\tau_{\rm XG} + \tau_{\rm GO} = \tau_{\rm XO}$ to a very high degree of accuracy. We therefore have that

$$I(\alpha) = \frac{f_{\rm X} D n_{\rm d}}{\sin\alpha} \int_\phi \frac{d\sigma}{d\Omega}(\phi, a) d\phi. \quad (6.46)$$

However, as we have already seen, interstellar grains have a size distribution $n_{\rm d}(a)da$ and this should be incorporated in Eq. (6.46); this is easily done by writing

$$I(\alpha) = \frac{f_{\rm X} D}{\sin\alpha} \int_\phi \int_a \frac{d\sigma}{d\Omega}(\phi, a) n_{\rm d}(a) da d\phi. \quad (6.47)$$

Eq. (6.47) can now be used to calculate the intensity of the scattered X-ray halo.

In practice, a major difficulty in detecting and measuring the X-ray haloes of the kind predicted by Eq. (6.47) is that the extent of the scattered halo ($\sim$ a few arcminutes; see above) is comparable, in the case of the *Einstein* imager, with the point spread function of the instrument. Nevertheless careful analysis of *Einstein* and *EXOSAT* images of suitable X-ray sources have revealed the existence of such haloes. An example is shown in Fig. 6.14. The point X-ray source GX $9+9$ is well out of the Galactic plane (its Galactic latitude $b = 9°$), where little interstellar dust is encountered along the line of sight. An X-ray halo is therefore not expected around this source and indeed, the X-ray intensity falls off with angular distance from the source in just the way expected for the point response function of the *Einstein* imager. The source GX 17 + 2, on the other hand, is close to the Galactic plane (its Galactic latitude $b = 1.3°$). In this case the variation of X-ray intensity with α is significantly above that expected from the point response function alone and indicates a substantial halo.

We shall see below that the observation of the polarization of starlight demands the existence of non-spherical grains, whereas the discussion thus far has been in the context of spherical grains, the easiest to treat analytically. We have already seen that the differential scattering cross-section for spherical grains is such that the scattering diagram can be described by a narrow cone in the direction of the incident X-ray. For non-spherical grains the corresponding cone is *least* extended in the direction in which the particle is *largest.* X-ray scattering therefore potentially presents a means of independently confirming the existence of non-spherical grains in the interstellar medium.

We finally note that there is a further effect at X-ray wavelengths that has potential in providing information about the chemical composition of interstellar dust grains. If an X-ray incident on an atom has energy $h\nu$ that exceeds the binding energy of electrons in the inner K shell, then the atom is ionized and the remaining electrons readjust their distribution amongst the energy levels. The ejected electron has kinetic energy $E = h\nu - \epsilon_K$, where ϵ_K is the binding energy of an electron in the K shell. In the non-relativistic case the electron momentum $p = \sqrt{2m_e E}$ and the electron has a de Broglie wavelength $\lambda_{deB} = h/p$. If the atom is free the electron de Broglie wave propagates away from its parent ion with spherical symmetry. In the case of a continuous X-ray source whose emission is observed through a cloud of identical atoms the X-ray spectrum would have an absorption edge–the K-edge–at photon energy ϵ_K. However if the atom is closely surrounded by other atoms (as in a solid) the electron de Broglie wave is scattered by neighbouring atoms and constructive and destructive interference of the de Broglie wave can occur. The result is that the absorption coefficient for a solid has considerable structure on the high energy side of the absorption edge, and this structure is characteristic not only of the ionized atom, but also of its nearest neighbours and their disposition. A detailed investigation of this structure (formerly called Kronig fine structure but now usually referred to as Extended X-ray Absorption Fine Structure–EXAFS) can therefore in principle provide information about the atomic environment of individual species in the solid phase. An example of EXAFS at the K-edge of silicon in silicon carbide is shown in Fig. 6.15. Note the sharp rise in the absorption coefficient at the K-edge, and the oscillatory structure (EXAFS) at higher energies.

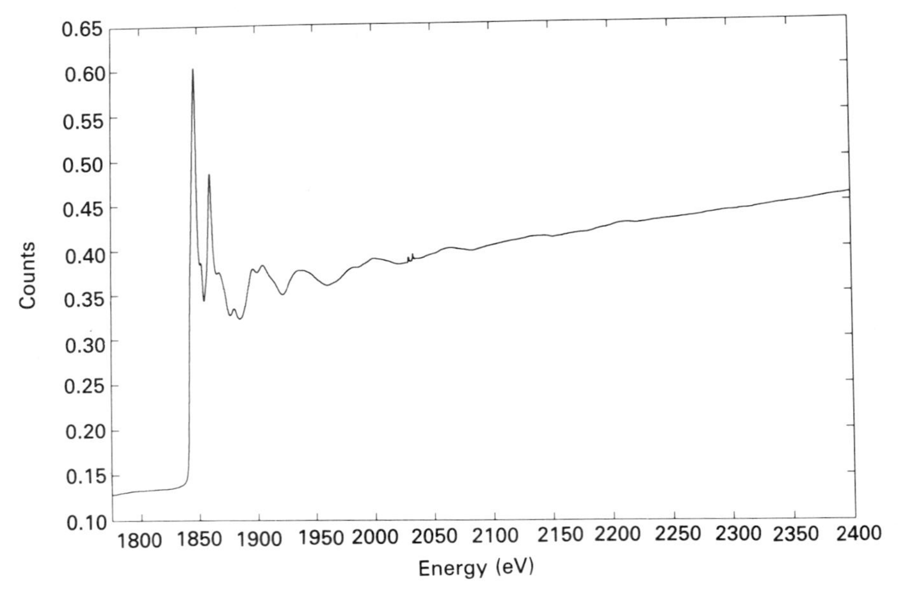

Figure 6.15: EXAFS at the K-edge of silicon in silicon carbide. Ordinate is absorption coefficient in arbitrary units, abscissa is X-ray energy in eV. After S. P. Thompson, Ph.D. thesis, Keele University (1991).

When high resolution X-ray spectrometers of sufficient sensitivity become available it may in principle be possible to use the EXAFS to determine the local environment of individual atoms in interstellar grains. One difficulty lies in the fact that a given atomic species (oxygen, say) will be present in both the solid (interstellar grain) and gas phases. The latter of course will not show the characteristic EXAFS: the absorption beyond the K-edge will be smooth and so the EXAFS arising from oxygen atoms tied up in grains will be 'diluted'. Ideally we require a species which is both cosmically abundant and which is also substantially depleted from the interstellar gas (see Table 6.2), and EXAFS from iron, carbon and oxygen may just be detectable in X-ray sources that are heavily reddened.

6.6 Emission by interstellar dust

6.6.1 The temperature of interstellar grains

In Section 4.3.1 we assumed that the power input into the grain is from a single star, assumed to be a point source. In some cases however this is not the approach required because the source of power surrounds the grain, sometimes entirely. A good example of this is the case of an interstellar grain, situated far away from any individual star and receiving radiation from general background starlight. For the moment we approximate the source of heating radiation by a blackbody, which emits radiation at a characteristic temperature T_{is}. In other words, we imagine that the grain is sitting at the centre of a cavity, the walls of which are emitting blackbody radiation at temperature T_{is}. Now a typical star has an effective temperature of a few thousand degrees, so let us take a value $T_{is} = 10^4$ K for definiteness. The energy density u_{BB} of radiation in a blackbody cavity is given by the laws of blackbody physics as aT_{is}^4, where a $(= 4\sigma/c)$ is the radiation constant; with $T_{is} = 10^4$ K, we have $u_{is} = 7.6$ J m^{-3}.

However an interstellar grain is clearly not situated in a blackbody cavity at 10^4 K: radiation with this characteristic temperature comes only from a few point sources (stars) scattered more-or-less at random over the celestial sphere. We must 'dilute' the blackbody cavity by a factor Ψ, so that the energy density of starlight in interstellar space $u_{is} = \Psi u_{BB}$. But the energy density of starlight can be directly measured, and is known to be $u_{is} = 7 \times 10^{-14}$ J m^{-3}. This gives us the value of Ψ as $u_{is}/u_{BB} \sim 10^{-14}$: the radiation field in interstellar space therefore resembles that in a blackbody cavity at 10^4 K, but diluted by $\sim 10^{-14}$. A more rigorous calculation might represent the radiation field in interstellar space as the sum of three or more blackbodies, each diluted by an amount $\sim 10^{-15}$–10^{-14}.

We can now estimate the temperature of a grain of radius a in interstellar space, again by equating the power going into the grain in the form of radiation $(4\pi a^2 \Psi \sigma T_{is}^4)$ to the power radiated when the grain has attained equilibrium $(4\pi a^2 \sigma T_d^4)$. This gives

$$T_d = \Psi^{1/4} T_{is}. \tag{6.48}$$

With $T_{is} = 10^4$ K and $\Psi \simeq 10^{-14}$ we have $T_d \simeq 3.2$ K for a blackbody interstellar grain heated by background starlight; again, in the blackbody case, the result is independent of grain radius. Now this temperature is close enough to that of the cosmic microwave background radiation ($T_{cos} = 2.735$ K) for us to wonder whether this might have a significant effect on the grain temperature. In this case the source of radiation is not, like the background starlight, diluted: this radiation is essentially isotropic and comes from the entire sky seen by the grain, so that $\Psi_{cos} = 1$ for this radiation. Proceeding as before we find

$$T_d = (\Psi_{is} T_{is}^4 + \Psi_{cos} T_{cos}^4)^{1/4} = 3.6K \tag{6.49}$$

so that the microwave background does, apparently, seem to have an effect on the temperature of an interstellar grain.

A more rigorous treatment would [like Eqs (4.14) and (4.15)] take into account the fact that the grains do not absorb and radiate like blackbodies: we must take into account their emissivity and its dependence on wavelength. The energy balance equation therefore reads

$$\Psi \int_0^\infty Q_{\mathrm{abs}}(\nu, a) B_\nu(T_*) d\nu = \int_0^\infty Q_{\mathrm{abs}}(\nu, a) B_\nu(T_{\mathrm{d}}) d\nu. \tag{6.50}$$

Using the Planck mean defined by Eq. (4.17), we can write Eq. (6.50) in the form

$$\Psi T_*^4 \overline{Q_{\mathrm{abs}}}(T_*, a) = T_{\mathrm{d}}^4 \overline{Q_{\mathrm{abs}}}(T_{\mathrm{d}}, a) \tag{6.51}$$

and, as we would expect from Eq. (4.21), the grain temperature depends both on the form of $\overline{Q_{\mathrm{abs}}}(T, a)$ and on the dimensions of the grain. We could also take into account the fact that the background radiation field is a combination of blackbodies, each with a specific dilution factor Ψ. Calculations along these lines indicate that interstellar carbon grains are expected to have temperature in the range 15 $\rightarrow$ 20 K for grains in the size range 0.01 $\rightarrow$ 0.5 μm, while silicate grains in the same size range have temperatures ranging from 14 to 18 K; in each case the smaller grains are hotter, as we would expect from Eq. (4.22).

We can use these results to calculate the expected infrared background intensity resulting from the emission of interstellar dust grains. Each individual grain emits like a blackbody at temperature T_{d}, with an emissivity given by $Q_{\mathrm{abs}}(\nu, a)$. Since both Q_{abs} and T_{d} depend on grain radius a we should sum over grain radii, using an appropriate grain size distribution. As usual, we use Eq. (2.11) with zero background term. The optical depth is given by

$$d\tau = \int_a Q_{\mathrm{ext}} n(a) \pi a^2 da ds,$$

and the source function is simply the Planck function:

$$S_\nu = B_\nu(T_{\mathrm{d}}).$$

In view of the fact that interstellar grains have temperature $T_{\mathrm{d}} \simeq 20$ K (see above), we might expect the corresponding infrared emission to resemble a Planck curve, weighted by the appropriate Q_{abs}; it will not look precisely like $Q_{\mathrm{abs}} B_\nu(T_{\mathrm{d}})$ because of the grain size distribution and the fact that the grain temperature is size-dependent, particularly for the smallest grains. The observed infrared background is shown in Fig. 6.16, in which it is evident that, at the longer infrared wavelengths ($\gtrsim 30\ \mu$m), the emission does indeed have the form expected. At shorter wavelengths, however, the intensity is many orders of magnitude higher than expected on the basis of a grain

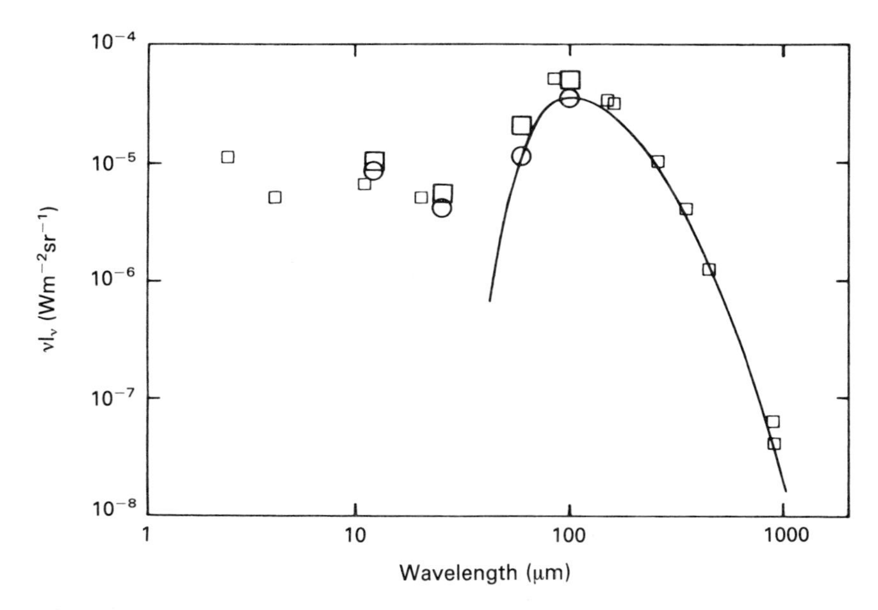

Figure 6.16: Infrared emission by interstellar grains. Curve is a fit based on infrared emission by grains heated by background starlight. After J. L. Puget & A. Léger. Reproduced, with permission, from the *Annual Review of Astronomy & Astrophysics*, Vol. **27**, p. 161. ©1989 by Annual Reviews Inc.

temperature $\simeq 30$ K. A possible reason for this discrepancy is discussed in Chapter 8.

6.7 Polarization by interstellar dust

Radiation from stars is almost entirely thermal in origin and as such, is not polarized. However, it is found that the light of many stars is indeed polarized, and that the degree of polarization is correlated both with location relative to the plane of the Galaxy and the amount of interstellar extinction and reddening.

In order to specify the linear polarization of a star we need two numbers, namely the *degree* of polarization and the *position angle.* The degree of polarization may be determined by using a filter which transmits electromagnetic radiation whose electric vector vibrates in a specific sense. The filter is rotated in the light path of the telescope until maximum intensity is obtained; this intensity is denoted by I_{max}. The process is repeated until minimum intensity (denoted by I_{min}) is obtained. The degree of polarization

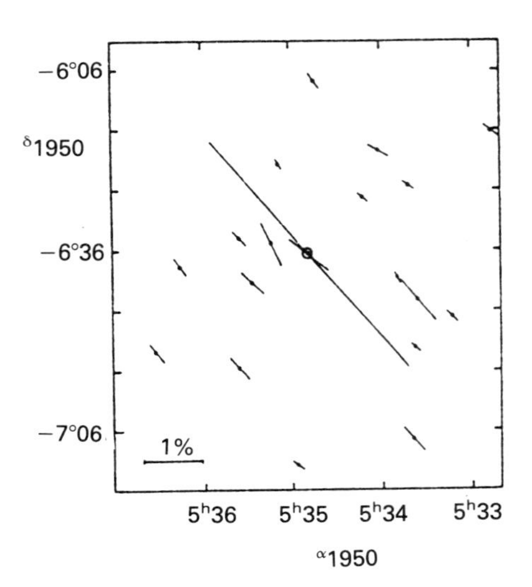

Figure 6.17: Polarization of a pre-main sequence star (BF Ori; near centre) and of background stars along the same line of sight. After V. P. Grinin et al., *Astrophysics & Space Science*, Vol. **186**, 283 (1991). Reprinted by permission of Kluwer Academic Publishers.

p is defined by (cf. Section 3.4.3)

$$p = \frac{I_{\max} - I_{\min}}{I_{\max} + I_{\min}}. \tag{6.52}$$

The orientation of the electric vector at maximum intensity (as determined by the orientation of the polarizing filter), projected onto the plane of the sky, defines the position angle of the polarization. It is customary to illustrate the polarization of starlight by plotting a straight line (the 'polarization vector') whose length is proportional to the degree of polarization and whose orientation is defined by the position angle. It may be that radiation from a star is intrinsically polarized, for example as a result of scattering of the stellar radiation by electrons or grains in the star's environment; in such a case the polarization of a number of stars, situated along the same line of sight to the star under study, is also measured so that the interstellar contribution to the polarization may be measured and removed (see e.g. in Fig. 6.17).

Also of interest is the *wavelength-dependence* of polarization, i.e. the degree of polarization at different wavelengths. It is found that the basic variation of polarization with wavelength is fairly standard from star

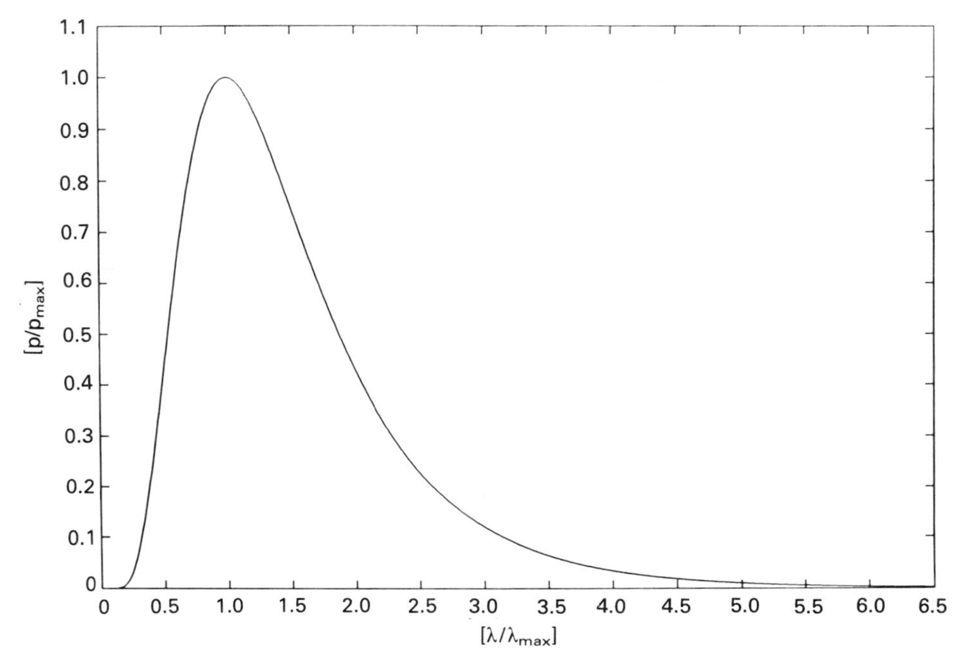

Figure 6.18: Wavelength-dependence of polarization.

to star, rising from short wavelengths, reaching a maximum value $p_{\rm max}$ at a wavelength $\lambda_{\rm max}$ and then falling to longer wavelengths (see Fig. 6.18). There are variations from star to star in the values of both $p_{\rm max}$ and $\lambda_{\rm max}$. However, if we normalize the polarization curves by plotting $p/p_{\rm max}$ against $\lambda/\lambda_{\rm max}$ instead of p against λ, we find that the polarization curves are virtually identical. The underlying curve is described by an empirical expression known as *Serkowski's law*:

$$\frac{p}{p_{\rm max}} = \exp\{-K(\log_e[\lambda_{\rm max}/\lambda])^2\}. \tag{6.53}$$

Earlier work suggested $K \simeq 1.15$ but more recent work, which has extended the polarization law into the ultraviolet and infrared, suggests a value of $K = -0.10 + 1.86\lambda_{\rm max}$ to take account of the fact that the $p(\lambda) - \lambda$ curve becomes increasingly narrow for increasing $\lambda_{\rm max}$. This curve is illustrated in Fig. 6.18. It is the wavelength-dependence of the polarization that holds the real key to understanding the origin of the polarization of starlight.

We have already seen in the discussion of reflection nebulae that scattered radiation is polarized in a sense that depends on the locations of the source of light, the scattering grain and the observer (see Fig. 3.11); however, this process is clearly not relevant in the case of interstellar polarization, which must have its origins in extinction rather than scattering.

Extinction by spherical grains will not preferentially absorb or scatter radiation having any specific state of polarization; non-spherical grains will however do so. Suppose we have an idealized situation in which we have a uniform distribution of cylindrical grains, each of radius a, and all with their long axes aligned parallel with each other. We consider the propagation and consequent extinction of a plane-polarized electromagnetic wave through this grain population. If the electromagnetic wave is polarized in such a way that the electric vector is parallel with the grain axes then the consequent extinction optical depth over a path length D is $\tau_{\mathrm{E}} = n_{\mathrm{g}} \pi a^2 [Q_{\mathrm{ext}}]_{\mathrm{E}} D$. For an electromagnetic wave which is polarized with its *magnetic* vector parallel with the grain axes the corresponding extinction optical depth is $\tau_{\mathrm{H}} = n_{\mathrm{g}} \pi a^2 [Q_{\mathrm{ext}}]_{\mathrm{H}} D$. In either case the efficiency factors $[Q_{\mathrm{ext}}]_{\mathrm{E}}$ and $[Q_{\mathrm{ext}}]_{\mathrm{H}}$ are as defined in Section 3.4.4.

In the astronomical case of course the incident radiation (starlight) is, in general, unpolarized. However, we can represent an unpolarized electromagnetic wave as a superposition of two incoherent polarized waves of equal amplitude, with electric vectors at right angles. Since $[Q_{\mathrm{ext}}]_{\mathrm{E}} \neq [Q_{\mathrm{ext}}]_{\mathrm{H}}$ we can expect a net polarization even though the original radiation was unpolarized. Suppose that, in the absence of extinction, the intensities would have been $I_{\mathrm{E}}^{(0)}$ and $I_{\mathrm{H}}^{(0)}$ for the two states of polarization, where $I_{\mathrm{E}}^{(0)} = I_{\mathrm{H}}^{(0)}$. In the presence of the cylindrical dust grains the observed intensities are

$$I_{\mathrm{E}} = I_{\mathrm{E}}^{(0)} \exp[-\tau_{\mathrm{E}}]$$

and

$$I_{\mathrm{H}} = I_{\mathrm{H}}^{(0)} \exp[-\tau_{\mathrm{H}}]$$

respectively. The *difference* in the extinction between the two states of polarization is

$$I_{\mathrm{E}} - I_{\mathrm{H}} = I_{\mathrm{E}}^{(0)} (\exp[-\tau_{\mathrm{E}}] - \exp[-\tau_{\mathrm{H}}]) \simeq I_{\mathrm{E}}^{(0)} [\tau_{\mathrm{H}} - \tau_{\mathrm{E}}];$$

the last approximation follows if both optical depths $\tau_{\mathrm{E}}, \tau_{\mathrm{H}} \ll 1$. Since $\tau_{\mathrm{E,H}} \propto [Q_{\mathrm{ext}}]_{\mathrm{E,H}}$ then

$$I_{\mathrm{E}} - I_{\mathrm{H}} \propto [Q_{\mathrm{ext}}]_{\mathrm{H}} - [Q_{\mathrm{ext}}]_{\mathrm{E}}.$$

Similarly the total extinction may be found by calculating the total intensity

$$I_{\mathrm{E}} + I_{\mathrm{H}} = I_{\mathrm{E}}^{(0)} (\exp[-\tau_{\mathrm{E}}] + \exp[-\tau_{\mathrm{H}}]) \simeq I_{\mathrm{E}}^{(0)} [\tau_{\mathrm{H}} + \tau_{\mathrm{E}}] \propto [Q_{\mathrm{ext}}]_{\mathrm{H}} + [Q_{\mathrm{ext}}]_{\mathrm{E}}.$$

Consequently the degree of polarization p is given by [cf. Eq. (6.52)]

$$p = \frac{[Q_{\mathrm{ext}}]_{\mathrm{H}} - [Q_{\mathrm{ext}}]_{\mathrm{E}}}{[Q_{\mathrm{ext}}]_{\mathrm{H}} + [Q_{\mathrm{ext}}]_{\mathrm{E}}}. \tag{6.54}$$

Note how the degree of polarization $p \propto [Q_{\mathrm{ext}}]_{\mathrm{H}} - [Q_{\mathrm{ext}}]_{\mathrm{E}}$; it is this property that enables us to use the polarization of starlight to estimate the sizes of interstellar grains. Suppose we have a star the light of which is polarized, the wavelength of maximum polarization $\lambda_{\mathrm{max}} \simeq 0.55\,\mu\mathrm{m}$. The variation of the efficiency factors $[Q_{\mathrm{ext}}]_{\mathrm{E}}$ and $[Q_{\mathrm{ext}}]_{\mathrm{H}}$ is such that $[Q_{\mathrm{ext}}]_{\mathrm{H}} - [Q_{\mathrm{ext}}]_{\mathrm{E}}$ peaks at a value $2\pi a/\lambda_{\mathrm{max}} \simeq 2$. This tells us immediately that $a \simeq 0.18\,\mu\mathrm{m}$; this is clearly consistent with the discussion of interstellar reddening in Section 6.4, in which the size of interstellar grains was estimated to be $\sim 0.1\,\mu\mathrm{m}$.

6.7.1 The alignment of interstellar grains

The fact that the polarization of starlight is a consequence of the unequal extinction for electromagnetic radiation having different states of polarization has led to a search for a mechanism that will cause interstellar grains to align themselves in a way that is consistent with observation.

In order to understand the way in which it is believed that grains are aligned in the interstellar medium we first look at a phenomenon that, at first sight, seems totally irrelevant to the discussion of cosmic dust. It has been known for some considerable time that there exists a Galactic radio background, due to synchrotron radiation emitted by highly relativistic electrons gyrating in the interstellar magnetic field. Synchrotron radiation is generally polarized, and for the simple case of an electron which has no motion along the field direction the polarization is linear and perpendicular to the direction of the magnetic field. Observation of the polarization of Galactic synchrotron radiation therefore effectively maps out the interstellar magnetic field. For lines of sight along which the polarization of starlight and of the Galactic synchrotron radiation can be directly compared there is strong evidence that the one is perpendicular to the other. Since we know that the polarization of the Galactic radio emission is physically connected with the orientation of the interstellar magnetic field the correlation between the optical and radio polarizations suggests strongly that the polarization of starlight is also in some way connected with the interstellar magnetic field. The most generally accepted theory along these lines was proposed as early as 1950 by Greenstein and Davis and, while there have been more recent variants, the essential idea behind the Davis-Greenstein mechanism remains valid.

We first note that a non-spherical grain immersed in the interstellar gas will acquire a rapid spinning motion as a result of random collisions with gas atoms. In thermal equilibrium we can expect that the energy associated with the grain's spin is comparable with the thermal energy of an atom in the gas; thus $\frac{1}{2}I\omega^2 \simeq kT_{\mathrm{gas}}$, where I is the moment of inertia of the grain about its axis of rotation and T_{gas} is the temperature of the gas. For a cylindrical grain of mass M and length $2l$, tumbling end-over-end, we have that $I = Ml^2/3$; we already know that $l \sim 0.1\,\mu\text{m}$ and $M \sim 6 \times 10^{-18}\,\text{kg}$. Since the interstellar gas is at temperature $T_{\text{gas}} \simeq 100\,\text{K}$, $\omega \simeq 3 \times 10^5\,\text{s}^{-1}$, in other words an elongated grain will spin rapidly (and randomly). The time taken by the grain to reach this state is approximately the time over which the grain collides with its own mass of gas. In the interstellar gas, with n_{H} hydrogen atoms per unit volume, the rate at which atoms strike a cylindrical grain, of length l and radius a, is $\sim n_{\text{H}} a l v_{\text{g}}\,\text{s}^{-1}$, where v_{g} is a typical thermal velocity for the gas atoms. The timescale for 'spinning up' the grain is therefore

$$\tau_{\text{spin}} \sim \frac{\pi a \rho}{m_{\text{H}} n_{\text{H}} v_{\text{g}}}.$$

For typical values of $v_{\text{g}} \simeq 1000\,\text{m}\,\text{s}^{-1}$ (corresponding to $T_{\text{gas}} = 100\,\text{K}$) and $\rho \simeq 2000\,\text{kg}\,\text{m}^{-3}$, we get $\tau_{\text{spin}} \sim 10^7$ years.

The essence of the Davis-Greenstein mechanism of grain alignment is an interstellar grain, containing paramagnetic impurities, spinning in the interstellar magnetic field; as a result of the rotation of the grain the internal field in the grain becomes misaligned with the external field, resulting in a torque which opposes any component of grain rotation which has an axis perpendicular to the interstellar magnetic field.

We consider a grain spinning with frequency $\omega \sim 10^5\,\text{s}^{-1}$ in the static interstellar magnetic field, the magnitude of which is $\mathbf{B_0}$; the static magnetic susceptibility of the grain is χ_0, i.e. in the absence of grain spin the magnetization $\mathbf{M} = \chi_0/(\mu_0\mu_{\text{r}})\mathbf{B_0}$, where $\mu_{\text{r}}(\simeq 1)$ is the relative permeability of the grain material; the field inside the grain is parallel with the external field. We have already seen that interstellar grains have temperatures in the region $10 \rightarrow 20\,\text{K}$, well below the Curie temperatures of most plausible grain materials (e.g. for iron, the Curie temperature is 1043 K). At these temperatures therefore the static susceptibility is given by Curie's law:

$$\chi_0 = \frac{\mu_0 \beta^2 g^2 N J(J+1)}{3kT_{\text{d}}}, \tag{6.55}$$

where β is the Bohr magneton, g is the Landé g–factor, N is the number of magnetic atoms per unit volume and J is the angular momentum quantum number; for Fe^{3+} ions, which we shall suppose to be responsible for the magnetic properties of the grains, $J = \frac{5}{2}$ and $g = 2$.

A spinning grain, however, does not experience a static field. It experiences a field whose magnitude varies as

$$B = B_0 \sin \omega t$$

and if ω is large enough the internal field lags the external field. This is expressed by writing the susceptibility as a complex quantity, $\chi = \chi' - i\chi''$. The real and imaginary parts of χ are, like the real and imaginary parts of the dielectric constant ϵ, related by Kramers-Kronig-type relations. Indeed there are obvious parallels here with the discussion of the interaction of electromagnetic radiation with a grain in Chapter 3.

At low frequencies, $\chi' \simeq \chi_0$ and $\chi'' = 0$, i.e. the particle behaves as though it were in a static B-field. The atomic dipoles in which the paramagnetism of the grain originate can 'keep up with' the changing B-field. At higher values of ω however, the dipoles fail to keep up with the external B-field and consequently they lag behind the rotation. The internal field therefore has a component that is perpendicular to the external field and a direction that tends to oppose the rotation of the grain.

For a grain rotating with frequency ω, the imaginary part of the susceptibility may be written as

$$\chi'' \simeq \chi_0 \frac{\omega}{\omega_0} \left(\frac{\pi}{2}\right)^{1/2}, \tag{6.56}$$

In Eq. (6.56) ω_0 is defined by the relation

$$\hbar\omega_0 = 2\beta^2\{8S(S+1)\alpha\}^{1/2}N,$$

where S is the total spin quantum number ($S = \frac{5}{2}$ for Fe^{3+}) and α is numerical constant that depends on how the magnetic ions are distributed through the grain.

By definition, the magnetization is the magnetic moment per unit volume; the magnetic moment of the grain is therefore $\mathbf{M}V$, where V is the grain volume. Since $\mathbf{M}$ and $\mathbf{B_0}$ are not parallel there is a retarding torque $V\mathbf{M} \times \mathbf{B_0}$ whose magnitude is

$$V|\mathbf{M} \times \mathbf{B_0}| \simeq KV|B|^2\omega,$$

where K absorbs all the constants in the magnetic moment, which acts to damp out any rotation about any axis that has a component at right angles to the field. This occurs on a timescale

$$\tau_{\mathrm{rel}} \simeq \left|\frac{\omega}{\dot{\omega}}\right| \tag{6.57}$$

$$= \frac{I}{KVB^2}, \tag{6.58}$$

where I is the moment of inertia of the grain.

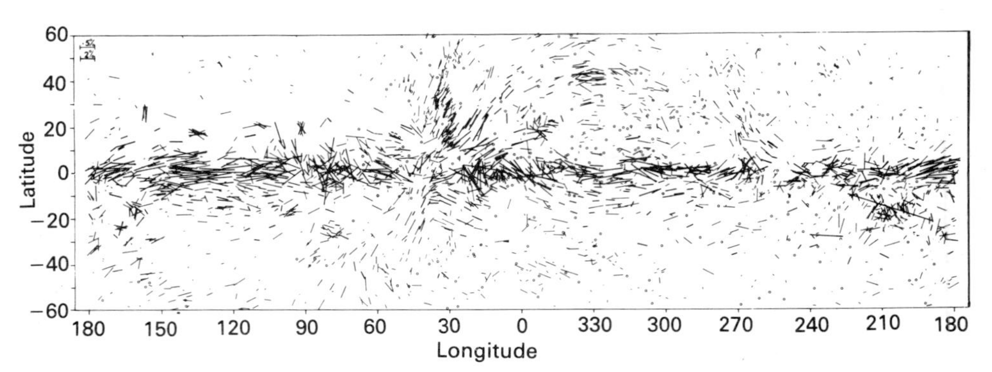

Figure 6.19: The distribution of interstellar polarization vectors on the plane of the sky. Co-ordinates are Galactic longitude and latitude. After D. S. Mathewson & V. L. Ford, ***Memoirs of the Royal Astronomical Society***, Vol. **74**, 139 (1970).

The net result is that grains, on the average, end up with their long axes perpendicular to the local magnetic field; we should therefore find that the polarization vectors of stars are well-organized on the sky when our line of sight cuts the aligning magnetic field at right angles, while the polarization vectors should be chaotically distributed when we view along the field. This is indeed what we find (see Fig. 6.19).

6.7.2 The relation between polarization and extinction

Since interstellar extinction and polarization have common origin in (presumably the same) interstellar dust grains, we might expect a relationship between the two. Although such relationships do exist they are somewhat complicated by the fact that, although extinction and polarization do have a common origin, the degree of polarization depends also on the degree of grain alignment and this is likely to vary from location to location. Furthermore, extinction over a given path length accumulates like a scalar quantity; however, the polarization of a star is defined by two numbers, the degree of polarization and the position angle, and the latter defines a direction on the plane of the sky. The polarization will therefore accumulate like a vector quantity so that polarization caused by propagation over part of the light path may be nullified by propagation over the remainder of the path.

In order to discuss the correlations that do exist we recall the dependence of polarization on wavelength, illustrated in Fig. 6.18, in which maximum polarization $p_{\max}$ occurs at wavelength $\lambda_{\max}$. While there is no tight correlation between $p_{\max}$ and $E(B-V)$ it turns out that $p_{\max} \leq 0.09E(B-V)$, the inequality occurring because of variations in the degree of grain alignment. A relationship is also found between optical-infrared reddening, as measured by

$$\frac{E(V-K)}{E(B-V)},$$

and the wavelength of maximum polarization λ_{max}; a recent determination gives

$$\frac{E(V-K)}{E(B-V)} = (1.01 \pm 0.44) + (3.80 \pm 0.71)\lambda_{max}; \tag{6.59}$$

there are similar, though less tightly-defined, relationships involving other colour excesses and λ_{max}.

As already noted the value of R, the ratio of total-to-selective extinction, differs in various directions and in different environments. There is a correlation between the ratio of total-to-selective extinction R and the wavelength of maximum polarization λ_{max}:

$$R = (-0.29 \pm 0.74) + (6.67 \pm 1.17)\lambda_{max}.$$

Note that, since $\lambda_{max} \propto a$ (see above), R can be used as a measure of grain size. This is especially useful in dark molecular clouds, where the extinction law can be determined both for embedded and background stars. Extrapolation of the extinction law to infinite wavelength (see Section 6.3) then gives the value of R, and hence of grain size, in the cloud. It is not uncommon for studies of this kind to show that the grains in molecular clouds are significantly larger than those in the general interstellar medium.

In many cases these relationships between extinction and polarization can be used to estimate the one if the other is known, although great care must be exercised in doing so. For example one has to be certain that there are no other contributions to the observed extinction or polarization other than those arising in the interstellar medium.

6.8 The formation of interstellar dust

If we refer back to Eq. (5.15), we can immediately show that interstellar grains can certainly not form in interstellar space. Suppose that n_X in Eq. (5.15) is independent of time. Then, assuming that $a \gg a_0$, the grain size after time t is

$$a(t) \simeq \frac{n_X S}{\rho} \left(\frac{kT_{gas} m_X}{2\pi} \right)^{1/2} t. \tag{6.60}$$

Rearranging Eq. (6.60) gives us the time $t(a)$ it takes for a grain to attain radius a:

$$t(a) = \frac{a\rho}{n_X S}\left(\frac{2\pi}{kT_{gas}m_X}\right)^{1/2}$$

[cf. Eq. (5.23)]. We have already seen that interstellar grains have dimensions that are typically $\sim 0.1\ \mu$m. The temperature of the interstellar gas is about 100 K and we can reasonably expect m_X to be about 20 m_H. Inserting these values in the above expression for $t(a)$ gives the time required for an interstellar grain to grow to the required size:

$$t \simeq \frac{2 \times 10^{12}}{n_X S} \text{ years} .$$

The number density of *hydrogen* in interstellar space is $n_H \sim 10^6 \rightarrow 10^7\ \mathrm{m}^{-3}$ while the number density of the *condensing species* is $\sim 10^{-4} n_H$ (cf. Table 6.2). We do not know the sticking probability accurately but we must have that $S < 1$ and so

$$t \gtrsim \frac{2 \times 10^{16}}{n_H} \text{ years} ,$$

considerably longer than the age of the Universe: clearly interstellar grains do not form in interstellar space. In order to form grains we need an environment in which the density is much higher, and an obvious place to look is in the environment of stars and in stellar atmospheres in particular. We return to this topic in Chapter 7.

We also note that Eq. (5.12) gives the rate at which atoms stick to the surface of a grain. This is also the rate at which atoms are removed from the gas phase so we might expect gas phase species to be *depleted* by an amount that is proportional to $m_X^{-1/2}$. However, we should not expect a correlation between the depletion (as given by Table 6.2) and the atomic weight of the depleted species in interstellar space because, as we have noted, interstellar dust grains do not form in interstellar space; such a correlation is more likely to arise in the winds of dust-forming stars.

6.9 Mantle formation

There are circumstances under which a grain, once formed, will acquire a mantle of material that differs from that of the original grain; for example ice (of whatever form) may form as mantles on silicate grains. Provided that the growth of such a mantle proceeds with spherical symmetry we may determine the rate at which the mantle is accreted in the same way that we derived Eq. (5.15). There is a slight difference however in that the species arriving on the grain surface need not necessarily be the same as the material from which the mantle is formed. For example, the growth of an ice mantle on a silicate grain in a molecular cloud can occur via a number of routes, such as the accretion of H_2O molecules directly from the gas phase, the accretion of the OH radical or the accretion of O and H separately. All these species are generally present in molecular clouds and the last two processes clearly involve the formation of H_2O by chemical reactions on the grain surface; these comments will of course apply to the formation of most ices. The rate at which the grain mass increases is therefore given by

$$\dot{m}_{\rm grain} = 4\pi a^2 n_{\rm X} m_{\rm man} S \left(\frac{kT_{\rm gas}}{2\pi m_{\rm X}}\right)^{1/2}, \tag{6.61}$$

where now we have to distinguish between the mass of the gas phase species ($m_{\rm X}$) and the molecular weight of the mantle material $m_{\rm man}$. The rate at which the grain grows is given by the equivalent of Eq. (5.14):

$$\dot{m}_{\rm grain} = \frac{4\pi a^2 \rho_{\rm man}}{3}\dot{a}, \tag{6.62}$$

where $\rho_{\rm man}$ is now the density of the mantle material. Combining Eqs. (6.61) and (6.62) as before we get

$$\dot{a} = \frac{3 n_{\rm X} m_{\rm man} S}{\rho_{\rm man}} \left(\frac{kT_{\rm gas}}{2\pi m_{\rm X}}\right)^{1/2}. \tag{6.63}$$

Stars within, or seen through, molecular clouds often show evidence of absorption features due to ices, such as water ice and CO. As already noted (Section 6.4.1) there is no evidence that water ice is present in the general interstellar medium but it does seem that conditions are favourable for the condensation of ices, including water ice, in molecular clouds.

The opacity to visible light in molecular clouds is generally high, and it is found that there is a threshold extinction before ice mantles begin to appear. This is probably due to the fact that, if short wavelength radiation can penetrate into the cloud, ices are sublimed or are prevented from forming in the first place. The value of the threshold can be determined by the following simple argument. Suppose we have a uniform spherical cloud, of radius $R_{\rm cloud}$, containing N_* stars per unit volume, of average luminosity L_* and effective temperature T_* (see Fig. 6.20). Consider a thin spherical shell, radius r and thickness dr, concentric with the cloud. The total luminosity of the stars contained in the shell is $4\pi r^2 dr N_* L_*$ and if the visual extinction

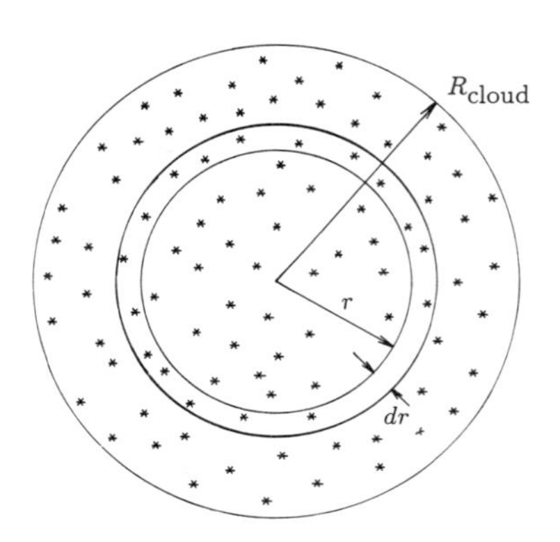

Figure 6.20: Estimating the threshold extinction for ice mantle formation.

coefficient for the cloud material is κ_V, the flux at the centre due to stars in the shell is

$$\frac{4\pi r^2 N_* L_* \exp[-\kappa_V r]dr}{4\pi r^2} = N_* L_* \exp[-\kappa_V r]dr.$$

The total flux at the centre, due to all the stars in the cloud, is obtained by integrating over all shells:

$$S = \int_0^{R_{\rm cloud}} N_* L_* \exp[-\kappa_V r]dr = \frac{N_* L_* R_{\rm cloud}}{\tau_V}(1 - \exp[-\tau_V]), \qquad (6.64)$$

where $\tau_V = \kappa_V R_{\rm cloud}$ is the visual optical depth in the cloud. We consider the energy balance for a grain at the centre of the cloud. The power input into a spherical grain, radius a, is $4\pi a^2 S\overline{Q_{\rm abs}}(T_*)$, while the power emitted by the grain is $4\pi a^2 \sigma T_{\rm d}^4 \overline{Q_{\rm abs}}(T_{\rm d})$. In equilibrium these two quantities are equal so that

$$\frac{N_* L_* R_{\rm cloud}}{\tau_V}(1 - \exp[-\tau_V])\overline{Q_{\rm abs}}(T_*) = \sigma T_{\rm d}^4 \overline{Q_{\rm abs}}(T_{\rm d}), \qquad (6.65)$$

where we have substituted for S from Eq. (6.64). In molecular clouds τ_V is usually large so we can safely set the term in round brackets in Eq. (6.65) equal to unity; also we shall approximate the Planck mean absorption efficiencies by $\overline{Q_{\rm abs}}(T) = AaT^\alpha$, where A and α are constants. If the grain temperature exceeds the evaporation temperature of the ice $T_{\rm evap}$, then no ice mantle can form; we must therefore have $T_{\rm d} < T_{\rm evap}$. Rearranging Eq. (6.65) we get

$$\tau_V > \frac{N_* L_* R_{\rm cloud}\overline{Q_{\rm abs}}(T_*)}{\sigma A a T_{\rm evap}^{(\alpha+4)}}. \qquad (6.66)$$

We see that the threshold extinction for displaying ice mantles on grains will depend on the nature of the ice (via T_{evap}) and the properties of the stars in the cloud (N_*, L_*).

Problems

6.1. The superluminal result in Section 6.5.3 appears to conflict with the results of the special relativity, which shows that c is a limiting velocity. How is this conflict resolved?

6.2. Estimate the time taken for a 0.1 μm silicate grain to grow a mantle of thickness 0.01 μm consisting of (i) carbon, (ii) CO ice. Assume a gas temperature of 20 K and hydrogen number density 10^9 atoms m^{-3} in each case. Note that CO ice might form by the direct accretion of CO from the gas phase, or of C and O separately, the CO in the latter case forming by reactions on the grain surface. Assume the 'solar' abundances of C and O relative to H given in Table 6.2, and that the gas phase CO abundance relative to H is 3×10^{-5}. The density of solid CO is $300\,\mathrm{kg\,m^{-3}}$.

6.3. If, as indicated in the previous problem, CO were to form on the surfaces of grains via the reaction of C and O (cf. the formation of H_2, Section 4.3.3), what effect would this have on grain temperature? The dissociation energy of CO is 11.1 eV.

6.4. Estimate the effect of the cosmic microwave background on the temperature of interstellar grains for the case where grains have absorption efficiency Q_{abs}.

Reading

The following, as the title implies, deals with a rather broader subject than that discussed in this book, but provides an excellent introduction to interstellar medium in general:

[C] *The Physics of the Interstellar Medium*, J. Dyson & D.A. Williams, Manchester University Press (1980).

Similarly, the following also deals with broader subject matter but is generally regarded as a classic work on the interstellar medium:

[D] *Physical Processes in the Interstellar Medium*, L. Spitzer, J. Wiley, Chichester (1978).

The following text provides a comprehensive discussion of interstellar dust, with much discussion of observational material, and ideally complements the material in this book:

[C/D] *Dust in the Galactic Environment*, D. C. B. Whittet, Institute of Physics (1992).

The topic of interstellar dust is reviewed in

[D] J. S. Mathis, *Annual Review of Astronomy & Astrophysics*, Vol. **28**, 37 (1990).

The paper by Herbig presents a classic discussion of the diffuse bands:

[D] G. Herbig, *Astrophysical Journal*, Vol. **196**, 129 (1975).

The following conference proceedings contain several review articles of interest:

[D] *Interstellar Dust*, Proceedings of International Astronomical Union Symposium 135, Eds L. Allamandola & A. G. G. M. Tielens, Kluwer Academic Publishers (1989).

[D] *Dust in the Universe*, Eds M. E. Bailey & D. A. Williams, Cambridge University Press (1988).

7

Circumstellar dust

7.1 Introduction

In this chapter we look at *circumstellar dust*–in other words dust grains that are closely associated with an individual star rather than being uniformly distributed through space. The evidence for dust grains being associated with stars comes from several quarters. It turns out that, in general, dust-bearing stars tend to be at extreme ends of the stellar evolutionary time-scale. Some are at an early stage of evolution, not yet at the stage (as is the Sun) where the star has settled into maturity; in these cases the dust tends to be either the remains of the material from which the star itself formed, or the result of dust formation in a strong 'stellar wind'. Others are highly evolved stars, often with chemical peculiarities, in whose atmospheres grain formation can occur. Indeed in some of the latter cases, the formation of dust can be seen 'in real time'.

7.2 What is circumstellar dust?

Before discussing circumstellar dust we should first establish what precisely we mean. In the last chapter we considered interstellar dust: what distinguishes 'circumstellar' dust from 'interstellar' dust? There are a number of ways of doing this. First, by definition, circumstellar dust is associated with an individual star, and its evolution and physical properties are dominated by the star. Thus, for example, its temperature is determined solely by radiative heating by the star and in general its temperature will be higher than that of a grain whose only source of heating is the background starlight of the interstellar radiation field. We can therefore define the extent of a circumstellar dust shell by determining the distance from the star at which the circumstellar grain temperature has dropped to the interstellar value.

As discussed in Section 6.6.1, the temperature of interstellar grains is about 20 K, while the temperature of a circumstellar dust grain is given by Eq. (4.21). In order to say something about the temperature of circumstellar grains we must say something about the nature of the grains (because of $\overline{Q_{\rm abs}}$) and of the central star ($L_{\rm bol}$). If we take carbon grains of 0.1 μm radius, the temperature of such grains is close to the interstellar value at a distance

$$r \simeq 3.7 \times 10^{16} \left(\frac{L_{\rm bol}}{L_\odot}\right)^{1/2} \text{ m}$$

from the star.

Another way is to look at the number density of dust grains in the circumstellar environment. In the general interstellar medium (i.e. not in dark clouds) there are $\sim 5 \times 10^{-7}$ grains m^{-3}. Near a dusty star there will be far more grains m^{-3} and we can again describe the extent of a circumstellar dust shell by specifying the distance form the star at which the number density of grains has declined to the general interstellar value. Again we need to specify some properties of the circumstellar dust before we can draw any conclusion. In particular we must decide on a density law for the circumstellar dust, i.e. the way in which the grain number density $n(r)$ varies with distance r from the star. It is often convenient to chose a power-law dependence on distance:

$$n(r) = n_0 \left(\frac{r}{r_0}\right)^{-\beta}, \tag{7.1}$$

where n_0 is the grain number density at some convenient reference distance r_0. The exponent β may be related to the origin of the grains; for example $\beta = 2$ if the grains form in a uniform outflow from the star (see below), while $\beta = 3/2$ for the dust shells of objects that are accreting from the interstellar medium, such as pre-main sequence stars. We shall assume the form (7.1) throughout.

From Eq. (7.1) we find that the total mass of circumstellar grains is

$$M_{\rm gr} = \frac{4\pi a^3 \rho}{3} \int_{R_{\rm in}}^{R_{\rm out}} 4\pi r^2 n(r) dr = \frac{16\pi a^3 \rho}{3} n_0 r_0^\beta \int_{R_{\rm in}}^{R_{\rm out}} r^{2-\beta} dr \tag{7.2}$$

if we have spherical symmetry. In Eq. (7.2) $R_{\rm in}$ and $R_{\rm out}$ are respectively the inner and outer radii of the dust envelope; frequently the value of $R_{\rm in}$ is determined by the condensation temperature of the dust. If we again take 0.1 μm carbon grains and suppose, for the purpose of getting an estimate, that $R_{\rm out} \gg R_{\rm in}$, we find that the number density of circumstellar grains has dropped to the interstellar value at a distance

$$r \simeq 3.3 \times 10^{15} \left(\frac{M_{\rm gr}}{10^{-6} M_{\odot}} \right)^{1/3} \text{ m.}$$

It seems then, that the properties of circumstellar dust might merge with those of interstellar dust at about 1 pc from the central star.

Finally, the very nature of circumstellar dust grains is different from that of interstellar grains. Their size, chemical composition etc. are determined, to a large extent, by the nature of the star with which they are associated, whereas the interstellar grain population is a mixture of all the grains ejected into the surrounding medium by the various types of dusty star, and further modified by prolonged exposure to the interstellar environment.

7.3 Stars with infrared excess

7.3.1 Infrared two colour diagrams

Infrared observations of certain stars reveal that they have circumstellar dust shells because they have an *infrared excess.* The concept of an infrared excess arises as follows. Suppose we have a star of known spectral type, so that we know from its optical spectrum what its temperature should be. This temperature determines the shape of its spectrum, from the ultraviolet, through the visible and into the infrared. If we know its flux at visual wavelengths we can extrapolate the shape of the spectrum into the infrared and determine what its infrared fluxes and colours should be. We can then measure the infrared fluxes and colours and compare them with those expected. It often happens that the measured infrared fluxes and colours tally well with those we would expect from the optical data; such stars would be of no interest in the present context. Of greater interest are those stars for which the fluxes *exceed* the expected fluxes; such stars are said to have an infrared excess, which can often (though not always) be attributed to emission by circumstellar dust. In such cases the infrared colours of the star will be redder (i.e. greater) than those expected on the basis of the optical data. In fact the colour temperature [e.g. as determined from an infrared colour index such as $(K - L)$] will be substantially cooler than that of the star itself. Examples of this are shown in Fig. 7.1, in which are shown the optical-infrared flux distributions of stars having circumstellar dust shells. Note that, as we pass out of the optical and go further into the infrared, the flux distribution falls away, roughly as λ^{-2}; this is simply the Rayleigh-Jeans tail of the stellar flux distribution. Longward of about 1 μm however, we see that the flux

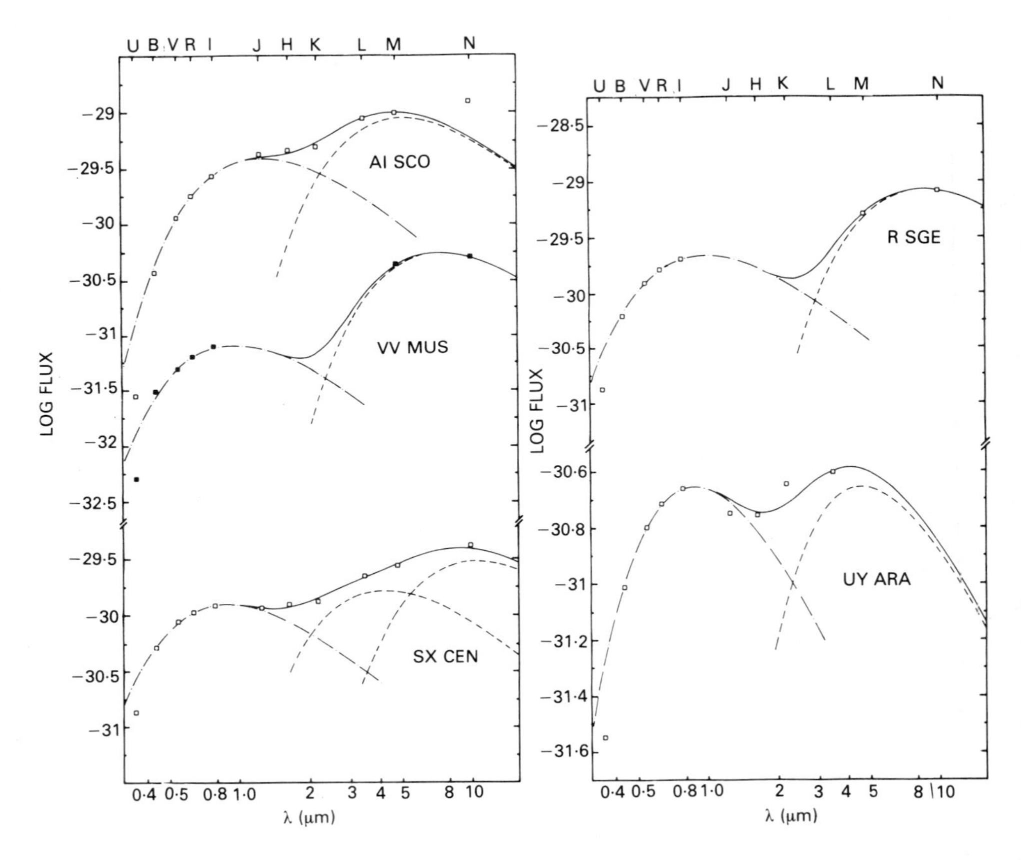

Figure 7.1: Infrared excess in stars with circumstellar dust shells. After M. J. Goldsmith et al., *Monthly Notices of the Royal Astronomical Society*, Vol. **227**, 143 (1987).

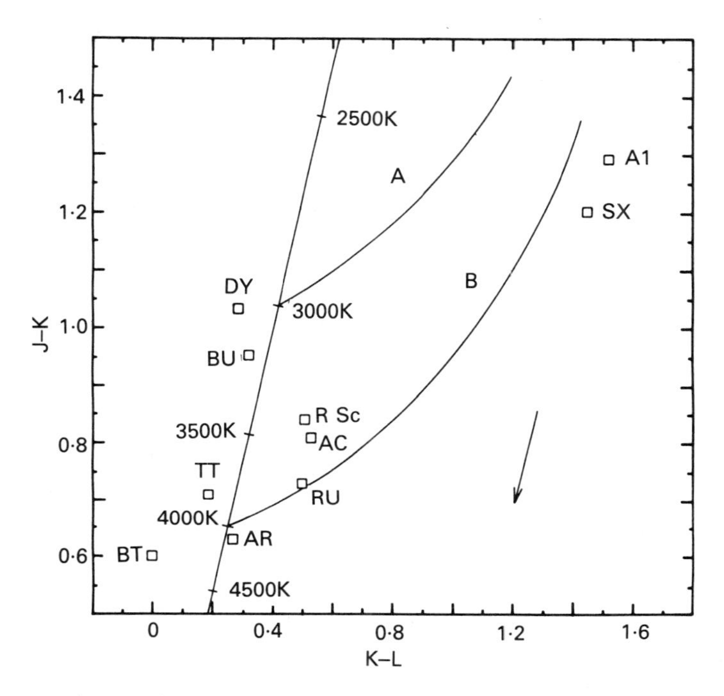

Figure 7.2: $(J-K)-(K-L)$ diagram. The near-vertical line corresponds to the locus of blackbody emitters; the tick marks correspond to the temperatures indicated. Curves A and B correspond to combinations of 3000 K (A) and 4000 K (B) emitters (stars) with 1000 K dust shells. Squares are data points. The arrow shows the effect of dereddening by an amount corresponding to $A_V = 1$ mag. After M. J. Goldsmith et al., *Monthly Notices of the Royal Astronomical Society*, Vol. **227**, 143 (1987).

distribution rises sharply, peaking at about 5 μm before falling away again at the longest wavelengths.

The infrared two colour infrared diagram can often be helpful in looking for evidence of an infrared excess; the $(J-K)-(K-L)$ diagram is illustrated in Fig. 7.2, which shows the locus of blackbody emitters having temperatures in the range 2500 $\rightarrow$ 4500 K. Most stars have effective temperatures $\gtrsim$ 2500 K so stars having no circumstellar dust envelope would be expected to lie close to this line. Curve A in Fig. 7.2 gives the locus of a combination of 3000 K and 1000 K blackbodies, the latter corresponding to emission by circumstellar dust. The location of an individual combination on this curve is determined by the relative contributions to the infrared flux from the star and its dust shell: the greater the contribution of the latter the higher the point on the curve. Curve B is for a 4000 K and 1000 K combination. A number of data points are plotted on Fig. 7.2. Those points clustered around the near-vertical line have no circumstellar dust shells (or at least none that show up in the near infrared); other points clustered around curve B show that the stars concerned have dust envelopes. Note that, even if there is no evidence of dust emission in the near infrared, emission in the far infrared by a cool dust envelope is not precluded. The existence of such dust shells may be deduced by constructing a far infrared two colour diagram, e.g. using *IRAS* fluxes.

Usually, of course, the circumstellar dust envelope reddens the star over and above any reddening imposed by *interstellar* dust. Furthermore it is almost certain that the reddening law for the circumstellar dust will differ substantially from that of interstellar dust. Fig. 7.2 shows the effect of interstellar reddening corresponding to $A_V = 1$. Note that the effect of interstellar reddening is to make a star appear cooler that it really is, with little or no effect on the presence or otherwise of an infrared excess.

7.3.2 Luminosity of the dust

We now estimate the total luminosity emitted by a circumstellar dust shell. Suppose we have a star of bolometric luminosity $L_{\rm bol}$, surrounded by a dust shell that completely surrounds the star, i.e. the dust shell subtends solid angle 4π at the star. Another way of expressing this is from the point of view of an observer situated on the star; from this observer's point of view the dust covers the entire sky.

An observer will see the star through its dust envelope and we can expect some extinction of the star's light. In order to proceed we make a couple of assumptions which may or may not always be appropriate. The first is an observational assumption: we suppose that the angular size of the dust shell is less than the size of the aperture being used to isolate the star, i.e. we assume that the star, together with its immediate environment, is a point source. The second concerns the scattering properties of the dust: we suppose that the dust grains scatter radiation isotropically. With these assumptions the extinction by the circumstellar dust arises as a result of *absorption* only; on the average, for every ray scattered out of the line of sight a ray is scattered in to the line of sight so that there is, effectively, no scattering. Thus the *observed* power from the star is

$$[L_{\rm bol}]_{\rm obs} = \int_0^\infty [L_\nu]_{\rm obs} d\nu = \int_0^\infty [L_\nu]_0 \exp[-\tau_{\rm abs}(\nu)] d\nu, \qquad (7.3)$$

where $[L_\nu]_0$ represents the intrinsic flux distribution of the star–i.e. in the absence of any circumstellar or interstellar extinction. Multiplying the top and bottom of the right hand side of Eq. (7.3) by $\int_0^\infty [L_\nu]_0 d\nu$ we can write

$$[L_{\rm bol}]_{\rm obs} = L_{\rm bol} \times \overline{\exp[-\tau_{\rm abs}]}, \tag{7.4}$$

where $\overline{\exp[-\tau_{\rm abs}]}$ is averaged over the flux distribution of the star, like the Planck mean of Eq. (4.17). [If the star does not radiate like a blackbody, or if the dust is not heated by a star at all–for example by the non-thermal radiation from the nucleus of an active galaxy–then $\overline{\exp[-\tau_{\rm abs}]}$ in Eq. (7.4) must be replaced by averaging over whatever flux distribution is appropriate.] Eq. (7.4) gives the *observed* bolometric flux from the star, and clearly $[L_{\rm bol}]_{\rm obs} < L_{\rm bol}$. The difference between $[L_{\rm bol}]_{\rm obs}$ and $L_{\rm bol}$ represents the stellar radiation absorbed by the dust, and which is re-radiated in the infrared. Thus

$$L_{\rm dust} = L_{\rm bol} - [L_{\rm bol}]_{\rm obs} = (1 - \overline{\exp[-\tau_{\rm abs}]}). \tag{7.5}$$

Alternatively,

$$L_{\rm dust} = [L_{\rm bol}]_{\rm obs}\{(\overline{\exp[-\tau_{\rm abs}]})^{-1} - 1\}. \tag{7.6}$$

Eqs. (7.5) and (7.6) are useful in getting first approximations to the extent of any circumstellar extinction. If we have a star with observed flux distribution S_λ, and good wavelength coverage in the optical and infrared, then a plot of λS_λ against λ (or of νS_ν against ν) shows the region of the spectrum in which the *power* (as opposed to the flux density) is emerging. The maximum value of λS_λ, denoted by $[\lambda S_\lambda]_{\rm max}$, gives a direct measure of the *bolometric* flux:

$$S_{\rm bol} \simeq 1.36[\lambda S_\lambda]_{\rm max}.$$

7.3.3 Flux distribution of circumstellar dust

In the simple case in which a circumstellar dust shell is optically thin, a distant observer will effectively see infrared emission from every dust grain in the shell and the total emission is obtained by summing the contribution from individual grains. We can get an idea of what the general dependence of the observed infrared flux distribution will look like from a fairly simple argument. We suppose that all the grains are identical, and have radius a and absorption efficiency $Q_{\rm abs}(\nu, a) \propto \nu^\alpha$ (cf. Section 3.4.2).

The distribution of grains in the shell will also contribute to the observed flux distribution. For example, if we have a uniform distribution of grains then the number of grains per unit volume between r and $r + dr$ does not depend on r (i.e. $n(r) \propto r^0$), whereas if we have grains forming in a steady stellar wind $n(r) \propto r^{-2}$ (see below). For the general case we write $n(r) \propto r^{-\beta}$, where β is a constant. There are therefore

$$4\pi r^2 n(r) dr$$

grains between r and $r+dr$. If the dust distribution is spherically symmetric, with the star at the centre, all the grains in the distance range $r \to r + dr$ from the star have the same temperature $T_{\rm d}(r)$, given by Eq. (4.22). Each grain will therefore emit at frequency ν at the rate $Q_{\rm abs}(\nu, a) B_\nu(T_{\rm d})$, where $T_{\rm d}$ is the grain temperature.

The total infrared luminosity of the shell is therefore

$$L_{\rm IR} \propto \int_{R_{\rm in}}^{R_{\rm out}} 4\pi r^2 n_0 (r/r_0)^{-\beta} Q_{\rm abs}(\nu) B_\nu(T_{\rm d}(r)) dr, \qquad (7.7)$$

where $R_{\rm in}$ and $R_{\rm out}$ are the inner and outer radii of the dust shell respectively. To proceed we write

$$\frac{h\nu}{kT_{\rm d}} = \frac{h\nu}{kA} r^{2/(\alpha+4)} = x,$$

where A is a combination of constants arising from Eq. (4.22); thus

$$dr = \left(\frac{\alpha+4}{2}\right) \left(\frac{kA}{h\nu}\right)^{(\alpha+4)/2} x^{(\alpha+2)/2} dx.$$

Substituting in Eq. (7.7), and distilling out the frequency-dependence by absorbing all the constants in the proportionality, we get

$$L_{\rm IR} \propto \nu^{(\beta-1)(\alpha+4)/2-1} \int_{x_1}^{x_2} \frac{x^{(\alpha+4)(3-\beta)/2-1} dx}{e^x - 1}, \qquad (7.8)$$

where x_1 and x_2 are the x values corresponding to $R_{\rm in}$ and $R_{\rm out}$. Now if the dust shell is (geometrically) thick enough that $x_1 \to 0$ and $x_2 \to \infty$, then the integral $\int_{x_1}^{x_2}$ approaches $\int_0^\infty$, a constant, and we find that

$$L_{\rm IR} \propto \nu^{(\beta-1)(\alpha+4)/2-1}; \qquad (7.9)$$

this shows that we expect to see a power-law dependence of the infrared emission on frequency.

Clearly the expression (7.9) can not apply over *all* frequencies, because in reality the shell has finite thickness: the grains occupy only the region $R_{\rm in} \le r \le R_{\rm out}$. The flux distribution must therefore turn over at low and at high frequencies. At the lowest frequencies the behaviour of $L_{\rm IR}$ must reflect that of the coolest grains present (i.e. those furthest from the star).

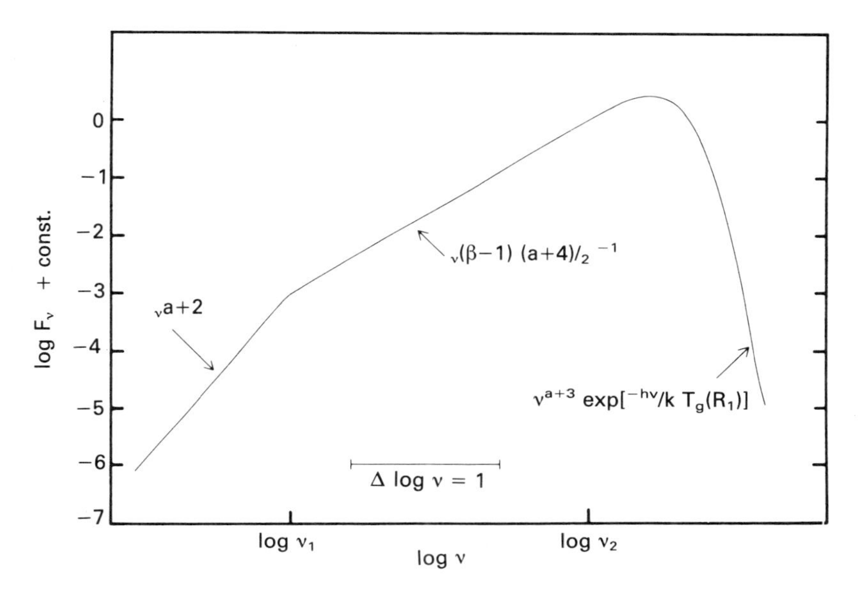

Figure 7.3: The infrared flux distribution from an optically thin dust shell. After M. F. Bode & A. Evans, ***Monthly Notices of the Royal Astronomical*** *Society*, Vol. **203**, 285 (1983).

At frequencies $\nu \lesssim [\nu_{\mathrm{max}}]_2$ therefore, where $[\nu_{\mathrm{max}}]_2$ is given by Eq. (1.20) with $T_{\mathrm{d}} = T_{\mathrm{d}}(R_{\mathrm{out}})$, we expect that

$$S_\nu \propto \nu^{(2+\alpha)}.$$

At the other extreme the flux distribution must reflect that of the hottest grains present, so that at high frequencies ($\nu \gtrsim [\nu_{\mathrm{max}}]_1$) we should find

$$S_\nu \propto \nu^{(\alpha+3)} \exp[-h\nu/kT_{\mathrm{d}}(R_{\mathrm{in}})],$$

where R_{in} is the inner dimensions of the dust shell. The expected flux distribution is illustrated in Fig. 7.3. Note that, at sufficiently long wavelengths (in practice, into the millimetre range), the shape of the spectral distribution depends only on α, which may therefore in principle be determined. Values of α determined in this way are generally in the range $1 \rightarrow 2$; a value at the lower end of the range would be typical of an amorphous conductor, whereas a value at the upper end is typical of crystalline materials.

7.4 Spectral signatures of circumstellar dust

Circumstellar dust may be detected in a number of ways. First we can look for an infrared excess as defined in Section 7.3; second, we can look for changes in the flux at short wavelengths resulting from variable extinction by circumstellar dust; third, we can use special techniques to directly image a dust shell, although in this case we need supplementary evidence to confirm the dusty nature of what we see.

There are several types of star which, by virtue of their infrared excess, are known to possess circumstellar dust shells. In such cases the dust may represent the remains of the material from which the star formed, or it may have formed as a result of condensation in the stellar environment. In the latter case, there is a subset of stars in which the formation of dust can be observed and followed in real time.

To a great extent the nature of any condensate is determined by the relative numbers of carbon and oxygen atoms in the environment in which grains condense. This is because the kinetic temperature in this region is $\sim 2000\,K$, at which the CO molecule is relatively easily formed and extremely stable. Thus whichever of O or C is underabundant relative to the other is likely to be depleted in the form of CO molecules, leaving whichever has the greater abundance to form grains. Thus if (as is usually the case) O is more abundant (by number) than C the carbon is locked up in CO molecules leaving the oxygen available for grain formation. Under these circumstances we would expect to see oxygen-bearing condensates, such as corundum (Al_2O_3), silica (SiO_2), magnesium oxide (MgO), magnetite (Fe_3O_4) and the silicates (e.g. olivine [Mg,Fe]SiO_3, enstatite $Mg_xFe_{1-x}SiO_4$). The precise nature of the condensate will also depend of course on the abundances of Al, Si etc. relative to each other and relative to oxygen in the grain-forming environment.

If, on the other hand, carbon is numerically more abundant than oxygen it is the oxygen that is tied up in CO molecules and the carbon is then available for grain formation. In this case the condensate will be carbon-bearing, such as carbon itself (e.g. in the form of graphite or amorphous carbon) or silicon carbide.

The nature of the environment in which grains condense will of course depend on the astrophysical nature of the underlying star. In many cases (e.g. supernovae) there will be substantial deviations from 'normal' cosmic abundance ratios because the nature of the star itself will involve thermonuclear processing which will change the initial relative abundances of the elements. Similarly, in environments in which *both* oxygen and carbon are underabundant, or in which the abundance of other species is anomalously

Table 7.1: Dust-related spectral features in circumstellar dust

Wavelength (μm)	Identification
0.250	Hydrocarbon molecules?
0.66	HAC
3.07	Amorphous water ice
3.28	PAH
3.4	PAH
3.53	Polyformaldehyde
6.2	PAH
7.7	PAH
8.6	PAH
9.7	Amorphous silicate
11.3	PAH
11.5	Silicon carbide
18	Amorphous silicate

high, we might expect more unusual condensates, such as MgS or Si_3N_4. Condensation temperatures for the more common condensates, calculated using the principles outlines in Section 5.4, are listed in Table 5.2, and have been calculated for typical cosmic abundances of the elements.

Many of the condensates mentioned above have spectral signatures which allow the identification of circumstellar dust. Most of these spectral features are in the infrared, and infrared spectroscopy has proved to be an indispensible tool in the identification of circumstellar dust grains. A listing of spectral features found in circumstellar dust shells is given in Table 7.1. Some of the features listed, such as the silicate feature at 9.7 μm, are commonly found whereas others, such as the water ice feature (3.07 μm) and a feature at 3.53 μm possibly associated with polyformaldehyde, are seen very rarely. The hydrocarbon and 'PAH' features are further discussed in Chapter 8.

The opacity of the circumstellar dust envelope will determine whether the spectral features summarized above are seen in emission or in absorption. A crude measure of whether the silicate or silicon carbide feature will be seen in emission or in absorption may be obtained by estimating the optical depth through the dust shell,

$$\tau = \int_{\textbf{pathlength}} \int_a Q_{\textbf{ext}} \pi a^2 n(r) da dr.$$

For those sources in which $\tau \lesssim 1$ we expect to see the feature in emission, while it will be in absorption otherwise. Fig. 7.4 shows a selection of spectra in the 7 → 23 μm wavelength range, obtained by the Low Resolution Spectrometer on the *IRAS* spacecraft.

7.4.1 Oxygen-rich giants

Stars which have evolved off the main sequence and which have entered the giant phase of their evolution are a major source of dust grains in the Galaxy. In general such stars have oxygen overabundant relative to carbon and therefore produce silicate dust.

All silicates consist of a tetrahedral network of SiO_2 in which Mg and Fe atoms are interspersed. Thus all magnesium and iron silicates, as well as SiO_2, have in common the Si–O bond, the stretching of which gives rise to a broad feature at 9.7 μm; thus we expect that all these materials will display this feature either in emission or in absorption. These materials also have a feature at 18 μm, resulting from the bending of the O–Si–O bond. Although the 9.7 μm feature is common to all silicates, the profile of the feature may provide further information about the precise nature of the silicate, e.g. whether amorphous or crystalline.

In modelling the silicate feature a number of possible approaches are possible. First one might take laboratory measurements of the optical constants of amorphous or crystalline silicates and attempt to match these with the observations. Suitable silicates can be 'manufactured' chemically and the resultant material, after drying, is ground down to produce grains of the required size. While the physical properties of such materials are easily characterized in the laboratory there must always be some concern that they may not match astrophysical silicates. The alternative therefore is to take an 'astronomical silicate', and the standard silicates of this kind are those seen in the infrared spectrum of the star θ Orionis (the so-called Trapezium), which typifies silicates around a young star, and those seen in the infrared spectrum of X Her, an evolved star. A compromise approach is to use the optical constants of solar system silicates, such as those recovered from primitive meteorites, or seen in the infrared spectra of comets. The profiles of the 9.7 μm silicate feature seen in various environments are shown in Fig. 7.5. Note the structure in the cometary silicate feature; this suggests that, unlike the amorphous silicate seen in the stellar environments, the cometary dust has a degree of crystallinity.

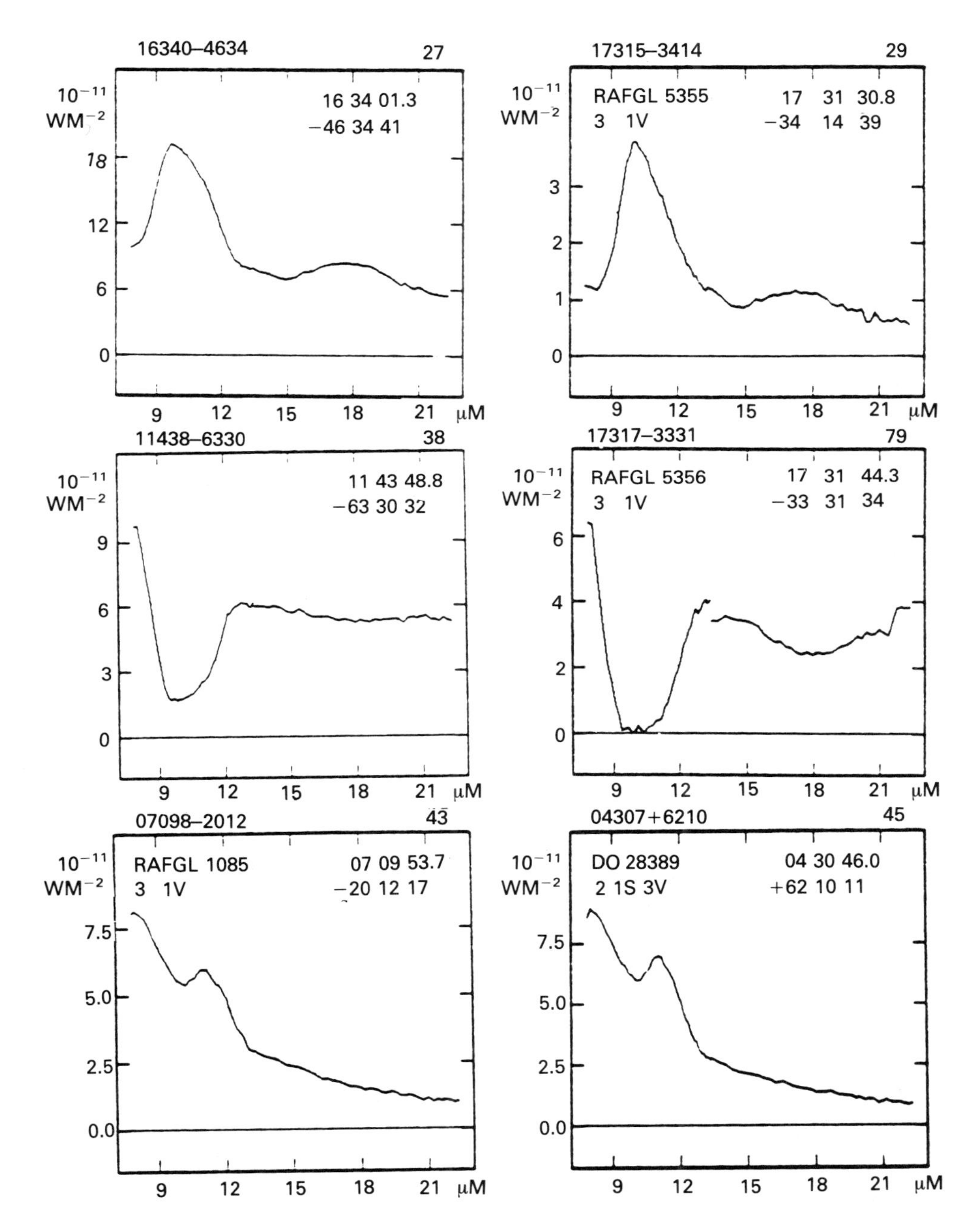

Figure 7.4: $7 \rightarrow 23\,\mu$m *IRAS* spectra. Upper two show the silicate $9.7\,\mu$m and $18\,\mu$m features in emission; middle two show the silicate features in absorption; lower two show silicon carbide emission. Number at top left of each frame is *IRAS* catalogue number; ordinate is λS_λ. After *IRAS* Science Team, *Astronomy & Astrophysics Supplement*, Vol. **65**, 607 (1986).

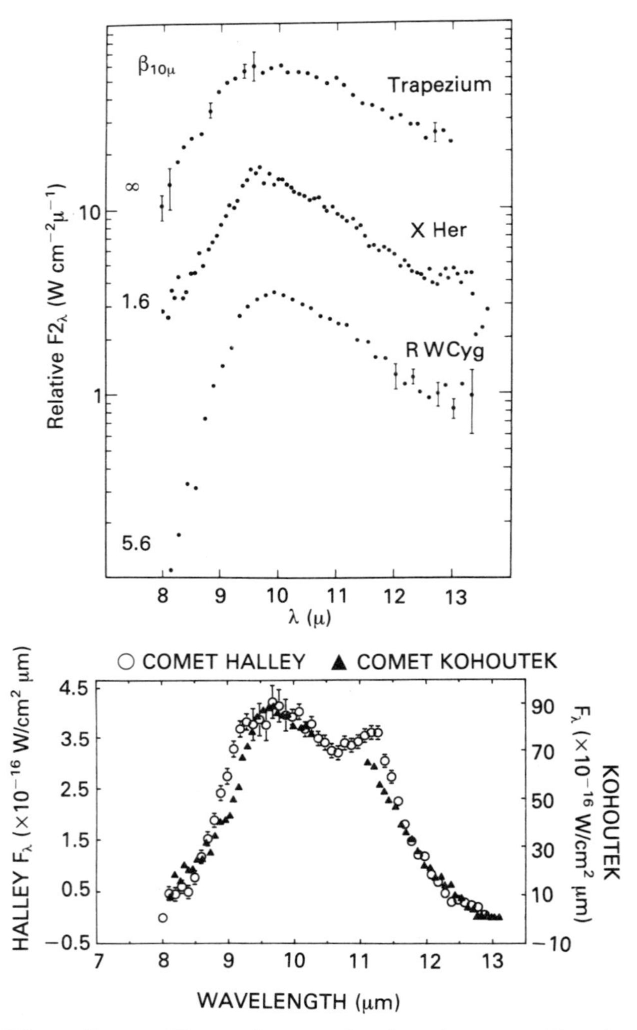

Figure 7.5: The 9.7 μm silicate feature in the circumstellar dust around the Trapezium stars in the Orion Nebula, X Her and RW Cyg (top frame), and in cometary dust (Halley and Kohoutek; lower frame). Stellar spectra after W. J. Forrest et al., *Astrophysical Journal*, Vol. **195**, 423 (1975); cometary spectra after H. Campins & E. V. Ryan, *Astrophysical Journal*, Vol. **341**, 1059 (1989).

7.4.2 Carbon stars

As stars undergo post-main sequence evolution there are phases in which their outer layers become carbon-rich. As discussed above these stars will form carbon or silicon carbide grains. The silicon carbide feature is commonly seen at wavelength $11.5\,\mu$m in the infrared spectra of carbon-rich stars. As in the case of the silicate feature the $11.5\,\mu$m feature is broad but unlike the silicate feature, has its origin in phonon modes in the SiC lattice; the minimum and maximum wavelengths of the feature correspond to the width of the forbidden gap in the SiC lattice dispersion relation (see Fig. 3.4).

Also seen in the infrared spectra of circumstellar dust shells are features associated with C–H bonding. For example, the stretching of the C–H bond in hydrocarbons, or on the surfaces of carbon grains, gives rise to features in the $3.2 \rightarrow 3.4\,\mu$m range, while the out-of-plane 'wagging' of the C–H bond in planar hydrocarbon molecules, whether in molecules or on grain surfaces, gives rise to a feature at $11.5\,\mu$m. The hydrocarbon features are further discussed in Chapter 8.

Searches for the circumstellar equivalents of the interstellar '2175Å' and diffuse features have not in general proved successful. There is a suggestion of an absorption feature at ~ 2500 Å in the spectra of some hot carbon-rich objects which has been attributed to very small ($\sim$ 50Å) carbon grains. The difficulty with searching for absorption features in this way is that one needs precise knowledge both of the intrinsic flux distribution of the star and of the wavelength-dependence of interstellar extinction along the line-of-sight to the star, neither of which is easily determined. A further difficulty with attributing such a feature to carbon grains of a specific size is that, in reality, we expect a grain size distribution and if this were the case, the agreement with observation is lost. Another possibility is absorption by hydrocarbon molecules, many of which have transitions around this wavelength.

7.4.3 Pre-main sequence stars

Pre-main sequence stars are stars which are at a very early stage of their evolution and are not yet, as their name implies, on the main sequence of the H–R diagram. Although the process of star formation is very poorly understood, it seems clear that it involves the collapse of a cloud followed by the formation of a protostar and an accretion disc, and the emission of material in the form of 'jets' at right angles to the disc. Such jets are often seen in the form of bipolar outflows, either at millimetre wavelengths by virtue of emission by molecules such as CO, or as scattered light (cf.

Section 6.5.4). As the star approaches the main sequence most–if not all– of the accreted material is dispersed. Pre-main sequence stars having mass similar to that of the Sun are known as T Tauri stars, after the brightest member of the class, whereas more massive stars are known as Herbig Ae or Be stars (the 'e' denoting that these are usually emission line stars).

By their very nature pre-main sequence stars are usually dusty and the circumstellar dust around these stars is detected by virtue of its infrared excess and in some cases, by variable extinction and polarization; indeed many of these stars are variable by virtue of varying amounts of obscuring material in the circumstellar environment between the star and the observer.

Fig. 7.6 shows the optical photometric and polarimetric variations of the Herbig Ae star WW Vul. There are clearly variations of large amplitude in the optical, and the nature of the photometric and polarimetric variations strongly suggest that whatever is responsible for the light variations also causes the polarimetric variations. Such variations can be understood in terms of clumps of material containing aligned grains orbiting the star; as a clump passes in front of the stellar disc it extinguishes the light from the star and at the same time causes a concurrent rise in polarization. The alignment of grains in these clumps may be due either to paramagnetic relaxation, as in the case of interstellar grains (see Section 6.7.1), or to gas streaming, in which the flow of gas forces elongated grains to lie with their long axes parallel with the flow of gas.

7.5 Grain formation and growth

7.5.1 The nucleation problem

The evidence that dust forms in stellar environments is conclusive–indeed as already discussed the formation and growth of dust grains is a phenomenon that can in some cases be followed in real time.

We would like to understand that process by which dust grains form in stellar atmospheres and in order to do so we need to know something about the nature of the grains. As we have already seen the grain temperature implied by the observations is in excess of 1000 K, so that we must be dealing with a refractory grain material: clearly, any ice, or even an ice derivative could not possibly survive at these temperatures. Generally speaking, we must confine ourselves to silicates, carbon, iron, silicon carbide or other similar material; as we have already seen the evidence in favour of silicates and silicon carbide from infrared spectroscopy is extremely strong. In discussing the formation of circumstellar dust, therefore, we ought to pay particular attention to these materials, although obviously we should ultimately keep

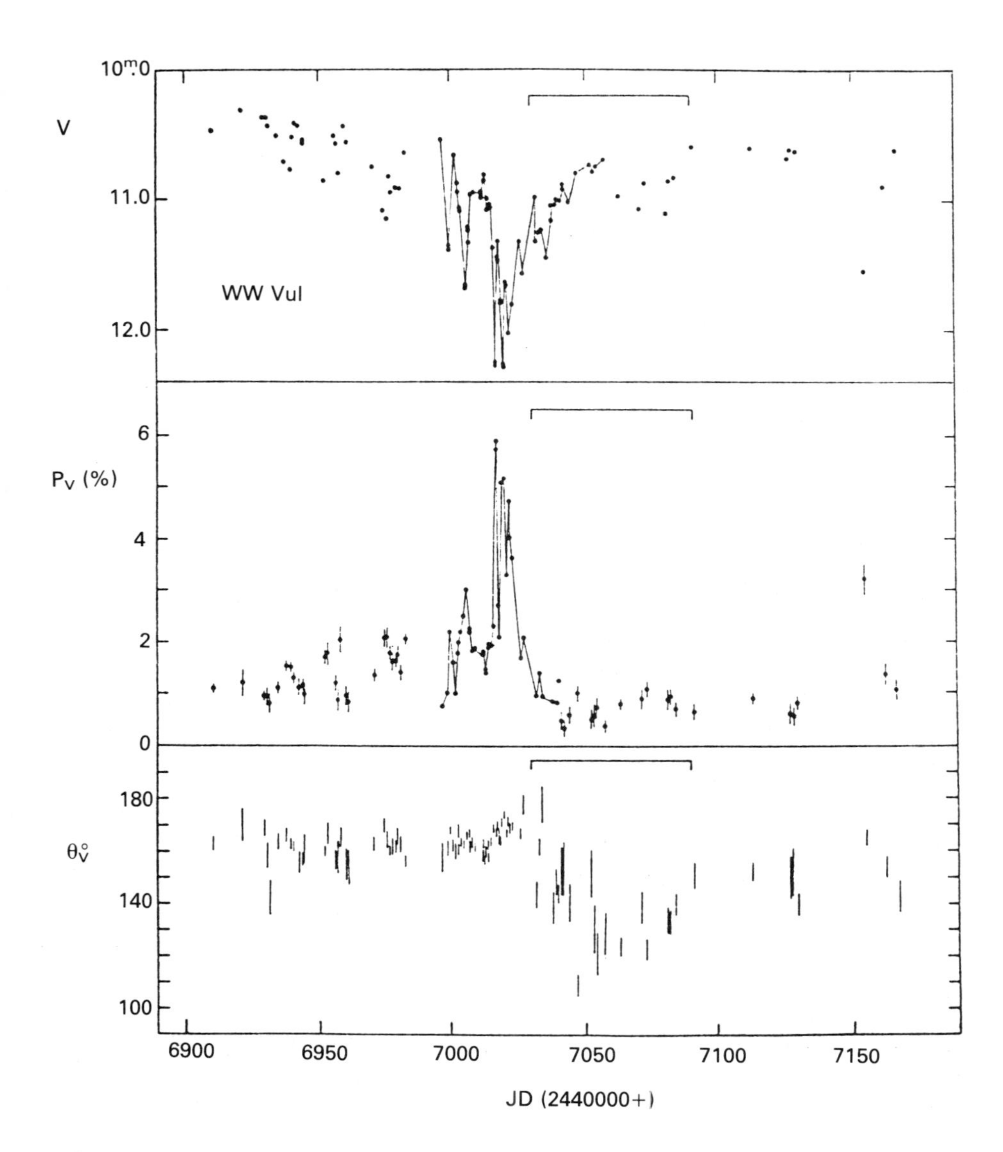

Figure 7.6: Photometric and polarimetric variations of the Herbig Ae star WW Vul. After V. P. Grinin et al., *Astrophysics & Space Science*, Vol. **186**, 283 (1991). Reprinted by permission of Kluwer Academic Publishers.

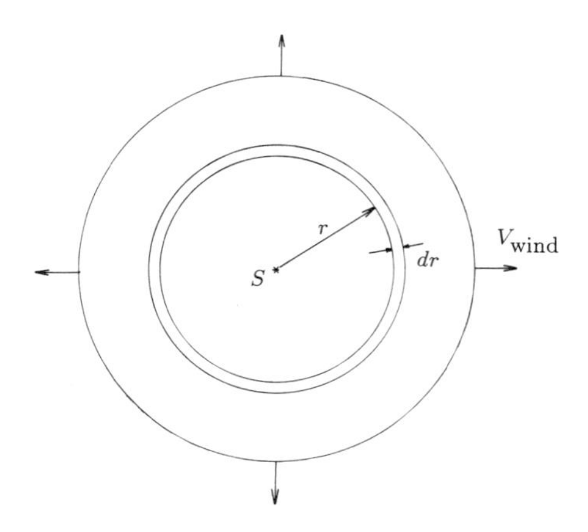

Figure 7.7: Calculating the density of gas in a stellar wind.

an open mind.

Basically, what we need to do is to understand how a material makes that transition from the gas phase to the solid phase. As already noted, once grain formation has got going in the first place, getting them to grow further is quite straightforward: it is the initial nucleation of the dust that poses a severe theoretical problem. What Nature seems to manage quite easily in stellar atmospheres is a process that has, so far, defied understanding as far as the basic physics of the process is concerned. As already noted (see Section 5.3), homogeneous nucleation is very unlikely and we must in general look at the possibilities of heterogeneous nucleation, possibly on ions in dust-forming stars of earlier spectral type but more likely on molecules.

7.5.2 Grain growth in a stellar wind

In a stellar wind matter is transported out from the stellar 'surface' into the surrounding environment. If we follow a parcel of gas in its outward journey from the star we will find that the number of atoms per unit volume in the parcel will decline. In the simple case of a uniform, isotropic, radial wind (i.e. there are no transverse motions) the decline can easily be calculated. We consider a thin spherical shell, of (constant) radius r and thickness dr; by virtue of the motion of the wind gas enters the inner surface of the shell and leaves the outer surface (see Fig. 7.7). Suppose that the wind results in mass-loss $\dot{M}$ from the star S, and that the wind has velocity V_{wind}. For most astronomical objects the bulk of the wind will be composed of hydrogen, for which the number of atoms per unit volume at distance r from the star is $n_{\rm H}(r)$. Since each atom has mass $m_{\rm H}$ the mass contained in the shell is

$$dM = 4\pi r^2 n_{\rm H}(r) m_{\rm H} dr.$$

Dividing each side by dt, and noting that $dr/dt = V_{\rm wind}$, we get

$$\frac{dM}{dt} = \dot{M} = 4\pi r^2 n_{\rm H}(r) m_{\rm H} V_{\rm wind}. \tag{7.10}$$

If we express the abundance (by number) of the condensate X as a fraction $f_{\rm X}$ of the hydrogen abundance, i.e. $n_{\rm X} = f_{\rm X} n_{\rm H}$, then

$$n_{\rm X}(r) = \frac{f_{\rm X}\dot{M}}{4\pi m_{\rm H} V_{\rm wind} r^2} \tag{7.11}$$

$$= \frac{f_{\rm X}\dot{M}}{4\pi m_{\rm H} V_{\rm wind}^3 t^2}, \tag{7.12}$$

where the second equality follows since $r = V_{\rm wind}t$. We see from Eq. (7.11) that, for a given parcel of gas, $n_{\rm X}$ declines with time as $n_{\rm X} \propto t^{-2}$.

Condensation of dust grains in the wind will only occur under suitable conditions. First, as we have already discussed, nucleation sites must be available; second, the temperature of any grains that do condense must be less than the condensation temperature of the material. The latter is equivalent to requiring that the partial pressure of the condensing species in the wind must exceed the saturation vapour pressure for the material. Clearly grain condensation in a stellar wind is not a foregone conclusion. Eq. (7.11) shows that the partial pressure of the condensing species ($\propto n_{\rm X}$ for constant gas temperature) in a given parcel of gas declines rapidly as it is carried away from the star by the wind; dust formation must therefore occur earlier rather than later, while $n_{\rm X}$ is sufficiently large. On the other hand the earlier condensation occurs the hotter any condensing grains will be, since grain temperature depends inversely on the distance of the grain from the star: cf. Eq. (4.13). Thus if $n_{\rm X}$ is only high enough when $T_{\rm d}$ is *too* high, grain formation will not occur at all. Even so grain formation in stellar winds is found to occur in a surprising variety of hostile environments.

Once nucleation has occurred, grain growth is comparatively easy. We consider a grain, of radius a, immersed in a gas at temperature $T_{\rm gas}$; for definiteness, we shall assume a carbon grain in a carbon vapour. Although a real astrophysical gas will contain other constituents (particularly hydrogen) it is with the partial pressure of carbon and the number density of carbon atoms with which we shall be concerned here. The rate of grain growth is given by Eq. (5.15) but in this case we have the complication that n_X depends on time by virtue of the expansion of the wind [Eq. (7.11)]. We therefore have

$$\frac{da}{dt} = \frac{\dot{M} S}{4\pi V_{\rm wind}^3 \rho t^2} \left(\frac{kT_{\rm gas}}{2\pi m_X} \right)^{1/2} . \tag{7.13}$$

We shall suppose in this case that it is only the expansion of the gas that contributes to the decline of n_X, in other words the depletion of the condensate is neglected.

Before we integrate Eq. (7.13) we must determine the limits of integration. As we have already noted, grain condensation is competing with the thinning out of the wind and with evaporation. For a given combination of star and grain type there is a specific distance–the *condensation distance*–at which grains will begin to condense; denote this distance by r_c. For a star of bolometric luminosity $L_{\rm bol}$ the condensation distance may be calculated from Eq. (4.21) by substituting $T_{\rm cond}$ for $T_{\rm d}$ and r_c for r and rearranging:

$$r_c = \left(\frac{L_{\rm bol}}{16\pi\sigma T_{\rm cond}^4} \frac{\overline{Q_{\rm abs}}(T_{\rm eff}, a_0)}{\overline{Q_{\rm abs}}(T_{\rm cond}, a_0)} \right)^{1/2} . \tag{7.14}$$

A parcel of gas transported out with the wind will reach this distance at time $t_c = r_c/V_{\rm wind}$. Prior to this time there are no grains, and we can suppose that growth begins at this time, with an initial size (as before) a_0. Solving Eq. (7.13) therefore gives

$$a(t) = a_0 + \frac{\dot{M} S}{4\pi V_{\rm wind}^3 \rho} \left(\frac{kT_{\rm gas}}{2\pi m_X} \right)^{1/2} \left(\frac{1}{t_c} - \frac{1}{t} \right) . \tag{7.15}$$

Note that, in this case, the rate of grain growth decreases with time; indeed, there is a maximum size to which the grain can grow, given by letting $t \to \infty$ in Eq. (7.15):

$$a_\infty = a_0 + \frac{\dot{M} S}{4\pi V_{\rm wind}^3 \rho} \left(\frac{kT_{\rm gas}}{2\pi m_X} \right)^{1/2} \frac{1}{t_c} . \tag{7.16}$$

In the wind of a carbon-rich red giant star of luminosity $10^3\, L_\odot$, we may find $V_{\rm wind} = 10\,{\rm km\,s^{-1}}$, while $\dot{M} = 10^{-6}\, M_\odot\,{\rm yr^{-1}} = 6.3 \times 10^{16}\,{\rm kg\,s^{-1}}$; of this only a fraction $\sim 10^{-3}$ is likely to be in the form of carbon. Carbon grains, of density $2200\,{\rm kg\,m^{-3}}$, would condense at distance 1.6×10^{11} m from the star, some 190 days after the grain-forming material had left the stellar surface. If the sticking probability is ~ 0.1 then substitution in Eq. (7.16) gives $a_\infty \simeq 0.6\,\mu$m. However if the condensing species is depleted the final size will of course be considerably smaller than this.

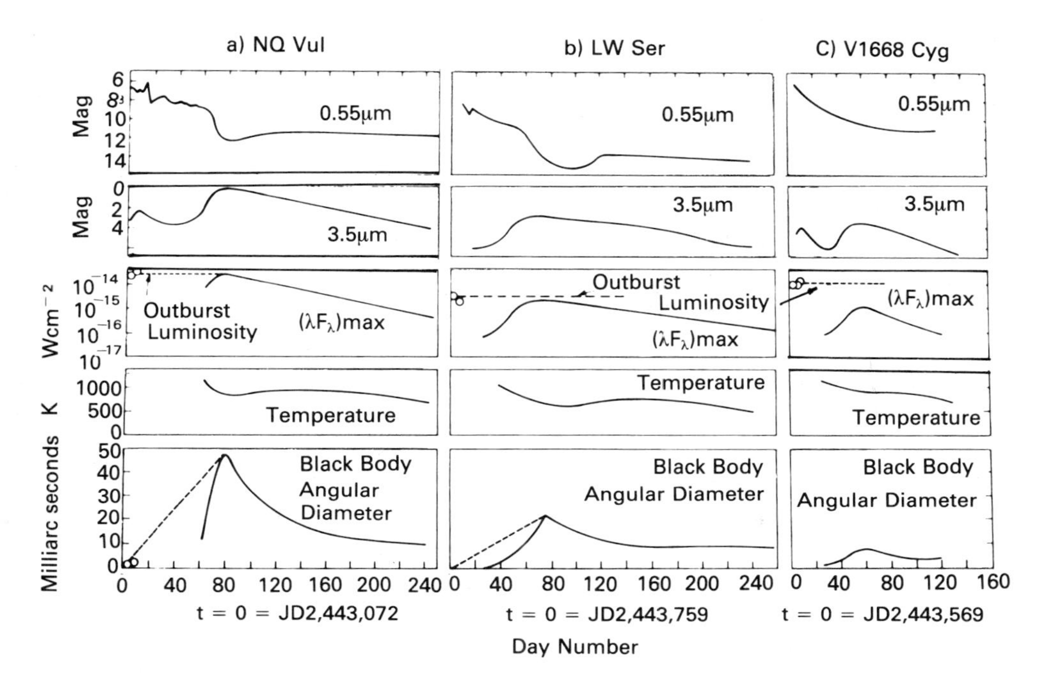

Figure 7.8: Visual and infrared light curves of novae NQ Vul, LW Ser and V1668 Cyg. Upper frame is visual light curve; second frame is infrared light curve; third frame represents infrared luminosity; fourth frame is grain temperature; fifth frame is size of emitting region. After R. D. Gehrz. Reproduced, with permission, from the ***Annual Review of Astronomy & Astrophysics***, Vol. **26**, 377. ©1988 by Annual Reviews Inc.

7.5.3 Stars undergoing sporadic grain formation

Novae

Novae are explosively variable stars in which the cause of the eruption is thermonuclear runaway on the surface of a white dwarf in a close binary system. The possibility that novae might be sites of dust formation dates back to the 1930's, and the eruption of one of the best studied of all novae, DQ Herculis (Nova Herculis 1934). More recently similar behaviour has been seen in a number of novae but in these cases, the optical observations are supplemented by observations at infrared wavelengths. The optical and infrared behaviour of three recent novae (NQ Vulpeculae 1976, LW Serpentis 1978 and V1668 Cygni) are illustrated in Fig. 7.8. In the case of DQ Herculis, two suggestion were made at the time to account for the deep minimum in the visual light curve. First, that molecules had formed in the material ejected in the nova explosion so that light from the underlying star was diminished; this hypothesis was quite attractive as CN molecules had in fact been detected in the optical spectrum of Nova Herculis shortly after maximum light. Second, that dust grains had formed in the ejected material. In reality, of course, molecules are not, like grains, capable of throwing a blanket over all wavelengths; even so, neither hypothesis was generally accepted and other explanations, not relevant to our present discussion, gained favour.

However, the infrared observations, typified by those illustrated in Fig. 7.8, clinch the argument for the dust-formation hypothesis. In the course of a nova eruption some 10^{26} kg of matter–overabundant in grain-forming material–is ejected into the surrounding medium. Once the ejected material gets to $\sim 10^{13}$ m from the centre of the explosion, conditions become suitable for the formation of dust grains. From the observer's point of view, this onset of grain formation is seen in two ways: first, the formation of dust renders the ejected material increasingly opaque, with the result that the light from the underlying star is diminished (as in Fig. 7.8) and second, the newly formed dust absorbs some of the radiation from the underlying star, and re-emits the absorbed energy in the infrared. Once the dust has dispersed, the visual light recovers before the final decline is resumed.

In the case of novae like NQ Vul, maximum infrared emission by the expanding dust shell occurs around the minimum in the visual light curve and at this time, the bolometric luminosity of the dust shell is comparable with the bolometric luminosity of the *stellar* remnant around maximum light. This confirms a prediction made by theories of the nova outburst: that despite the decline in the visual, the *bolometric* luminosity of the nova remnant remains constant while the outburst progresses.

An interesting property of nova dust is related to the way in which the dust-forming material has been produced. The thermonuclear reactions that give rise to the eruption will imprint on the ejected material a specific isotopic signature, which is seen in novae and not elsewhere. Nova dust is therefore likely to be isotopically labelled as such and indeed, it has been suggested that some isotopic anomalies in meteorites may be explained in this way. Fig. 7.9 shows the isotopic ratios $^{15}N/^{14}N$ determined by the analysis of meteorites, plotted against $^{13}C/^{12}C$ from the same source. Note that each source of dust (such as novae, evolved giants etc.) has a readily identifiable isotopic signature, suggesting that the origin of some of the material in meteorites may be traced back to specific stellar sources.

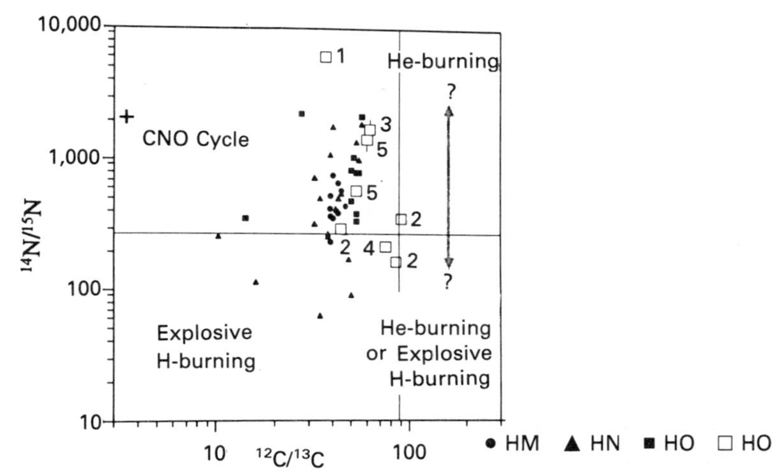

Figure 7.9: Isotopic ratios in meteorites. Materials occupy different regions of the diagram depending on the thermonuclear origin of the N and C, as indicated. After Tang Ming et al. Reprinted with permission from *Nature*, Vol. **339**, 351 ©1989 MacMillan Magazines Ltd.

Supernovae

Like novae, supernovae are explosively variable stars but in this case, the eruption represents the complete disintegration of the star, generally leaving a neutron star or black hole remnant. *Type I* supernovae occur in binary systems, while *Type II* supernovae result from the collapse of the core of a massive ($M \gtrsim 8\,M_\odot$) red supergiant star. We are generally handicapped in the study of supernovae by the fact that, with the notable exception of SN1987A, we are generally confined to observing them in distant galaxies; the result is that, despite their great luminosity ($\sim 10^9\,L_\odot$), supernovae are extremely faint.

As was the case with novae, the suggestion that dust grains might form in the debris resulting from a supernova explosion is not a new one. Material ejected in a supernova explosion travels at some $10^4\,\mathrm{km\,s^{-1}}$, while the peak luminosity of the supernova is typically $10^{36}\,\mathrm{W}$. If we assume blackbody grains to get orders of magnitude, so that the $\overline{Q_{\rm abs}}$ factors in Eq. (7.14) are unity, then we can easily estimate the condensation distance $r_{\rm c}$ if the condensation temperature is $\sim 1500\,\mathrm{K}$. Since $t_{\rm c} = V\,r_{\rm c}$ we see that dust is unlikely to condense in supernova ejecta before a year has elapsed, and by this time after outburst supernovae in distant galaxies become extremely faint. This is not to say, however, that there is no evidence for dust around supernovae; the first infrared excess associated with a supernova was discovered in the case of SN 1979C in the galaxy M100 and several more instances have been discovered since. However, the dust grains around these supernovae seem not to have condensed from the ejected material. Instead the infrared emission that is seen arises from the heating of a pre-outburst dust shell, which originated as a result of grain condensation in the wind of the progenitor star.

The dust shell around a potential supernova will be quite extensive, so that light will take considerable time to travel from the erupting supernova out to the surrounding dust grains. The result is that we get an 'infrared echo'–the infrared equivalent of the light echo discussed in Section 6.5.3. The only supernova in which grain condensation in the ejected material has been unambiguously detected is SN 1987A in the Large Magellanic Cloud. We note that supernova dust, like nova dust, will bear unique elemental and isotopic signatures.

R CrB stars

As we have already mentioned, most astronomical objects contain some 73 per cent hydrogen by mass, with 25 per cent helium and 2 per cent heavy elements. However, there is a remarkable class of star in which this rule of thumb is not even remotely correct; these are the *hydrogen-deficient stars* in which, as their name suggests, hydrogen may be a very minor constituent. There is a sub-class of the hydrogen-deficient stars in which carbon is over-abundant; these are the R Coronae Borealis, or RCB, stars, named after the 6 th magnitude proto-type.

The light curve of the RCB star RY Sagittarii is shown in Fig. 7.10. Most of the time the star may be seen around sixth magnitude, the slight variations that do occur being due to the pulsation of the star itself. The most dramatic features of the light curve, however, are the unpredictable drops in light, often by as much as 8 magnitudes. This behaviour was discovered in 1793 by Piggott, who did much of the pioneering observational work on stellar variability. This behaviour was explained independently in the 1930's by Loreta and O'Keefe. These workers were aware of the chemical peculiarity of RCB stars and suggested that their bizarre photometric behaviour could be understood in terms of clouds of carbon dust, ejected by the star in the direction of the observer. This explanation–the 'Loreta-O'Keefe hypothesis'– is now generally accepted as the correct explanation of the behaviour of the light curve of RCB stars. In recent years, infrared observations of RCB stars during decline have demonstrated that there is no increase in the infrared flux as the visual light curve goes into decline: it seems clear that, at infrared wavelengths, there is negligible additional emission by the condensing dust. This contrasts with novae (see above), in which material is ejected in all

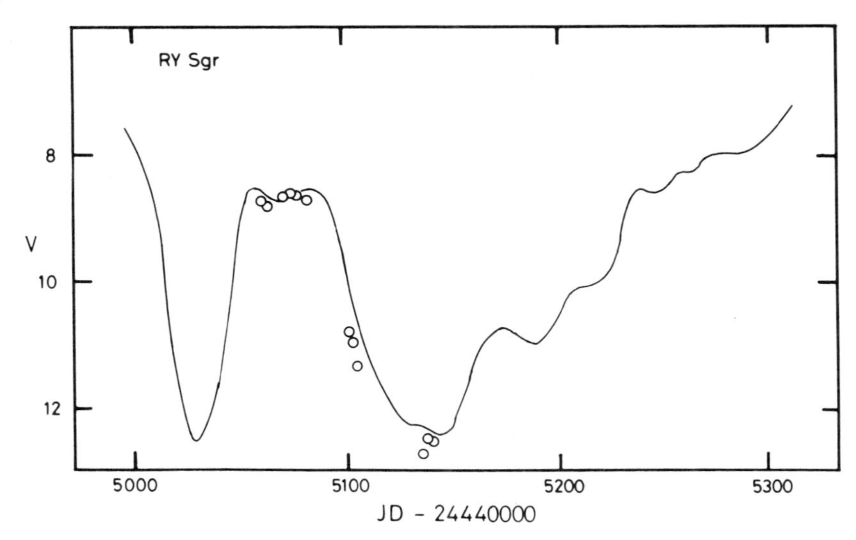

Figure 7.10: Visual light curve of the RCB variable RY Sagittarii during a deep minimum. After A. Evans et al., *Monthly Notices of the Royal Astronomical Society*, Vol. **217**, 767 (1985).

directions. In the case of the RCB stars their behaviour lends support to the view that the emission of material takes place only over a limited region of the stellar surface.

Problems

7.1. In deriving Eq. (7.5) it was assumed that the circumstellar dust scatters isotropically and that the circumstellar envelope is contained within the telescope beam. How would Eq. (7.5) be modified if either or both of these assumptions did not hold?

7.2. Show that $S = 1.36[\lambda S_\lambda]_{\rm max}$.

7.3. Calculate the condensation distance $r_{\rm c}$ and condensation time $t_{\rm c}$ for (i) a nova having $L_{\rm bol} = 10^{31}$ W and $V_{\rm wind} = 800\,{\rm km\,s^{-1}}$ and (ii) a supernova having $L_{\rm bol} = 10^{36}$ W and $V_{\rm wind} = 10^4\,{\rm km\,s^{-1}}$. Assume $T_{\rm cond} = 1500$ K in both cases and that $\overline{Q_{\rm abs}} = 10^{-6}\,a\,T^2$ (a in μm).

Reading

The following references contain several reviews on circumstellar dust; the first includes discussions of the rôle of dust grains in the early and late phases of stellar evolution, while the second contains discussions of the connection between circumstellar and interstellar dust:

[D] *Circumstellar Matter*, Proceedings of International Astronomical Union Symposium 122, Eds I. Appenzeller & C. Jordan, D. Reidel Publishing Company (1987).

[D] *Interstellar Dust*, Proceedings of International Astronomical Union Symposium 135, Eds L. J. Allamandola & A. G. G. M. Tielens, Kluwer Academic Publishers (1989).

8

The PAH hypothesis

8.1 Introduction–How small can a solid be?

We have already seen that the dust particles in interstellar and circumstellar space are very small, certainly no larger than $\sim \mu$m in size. A simple calculation demonstrates that a typical grain can therefore contain only a few $\times 10^6$ atoms at most, while the smallest grains of all–such as the ones that will suffer stochastic heating–contain considerably fewer than this. This smallness of cosmic dust grains obviously begs the question 'How small can a cluster of atoms be and still have the properties of a solid?' or, equivalently, 'How large can a molecule be and still retain the properties of a molecule?' In other words, what distinguishes a 'large molecule' from a 'small (solid) dust grain'? A further problem, which those who are interested in determining the laboratory properties of cosmic dust grains must address, is whether the properties of solids measured in bulk in the laboratory, have any relevance to the properties of the small sub-μm particles in interstellar and circumstellar space.

To some extent, the properties of a substance (whether 'molecular' or 'solid') depend on the property in which we are interested. For example, we know that to eject an electron from a molecule requires an energy at least equal to the ionization potential of the molecule, whereas to eject an electron from the surface of a solid requires an amount at least equal to the work function of the solid. In order to determine the critical size at which a 'molecule' makes the transition to a 'solid' we could measure the ionization potential of clusters of atoms of increasing size and determine the cluster size at which the ionization potential approaches the work function.

Another criterion would be to look at the dissociation energy of increasingly large clusters of atoms and (as before) determine the cluster size at which the dissociation energy approaches the heat vaporization per atom of the corresponding solid.

Yet another measure of whether or not a cluster of atoms behaves like a molecule or a solid is the existence of discrete energy levels or of energy bands; in the former case we are probably justified in regarding the collection as a molecule and in the latter case, the cluster behaves like a solid. For some applications the energy levels of a collection of atoms are likely to be in the form of bands rather than discrete levels if an atom has nearest and next-nearest neighbours. The 'neighbours' criterion is also one that determines whether or not the collection of atoms has a 'surface', which is another solid-state rather than molecular concept. In practice this means that a collection of at least ~ 100 atoms grouped into a volume ~ 10 Å in dimensions is likely to have band structure and can be regarded as a solid. It is with the properties of small clusters of atoms that we shall be concerned in this chapter.

8.2 Anomalies in the infrared emission of interstellar dust

One of the surprises of the *IRAS* mission was the discovery of the so-called infrared cirrus (see Fig. 8.1). This emission is almost certainly due to interstellar grains, although atomic and ionic emission processes have also been suggested. Contrary to what one would expect from the disc-like shape of the Galaxy and an assumed uniform dust distribution, the cirrus emission is extremely patchy and extends to high Galactic latitude.

Given the composition and size distribution of interstellar grains it is possible to calculate their temperature and hence infrared emission. As reliable far infrared data became available it became apparent that there is a serious discrepancy between the predicted and observed infrared emission in the sense that the latter exceeds the former by several orders of magnitude at wavelengths $\lesssim 30\,\mu$m. This is well illustrated in Fig. 6.16, which shows that, whereas the emission for wavelengths longer than about $30\,\mu$m seems reasonably consistent with that expected from grains at temperature 20 K (consistent with the discussion of Section 6.6.1), there is a large excess at shorter wavelengths.

There are similar discrepancies in the infrared flux distributions of individual objects, such as H II regions and reflection nebulae. Indeed, in many cases there is evidence that the colour temperature of the dust does not, as would the *equilibrium* grain temperature predicted by Eq. (4.13) or (4.22), decrease with increasing distance from the central star, suggesting that the

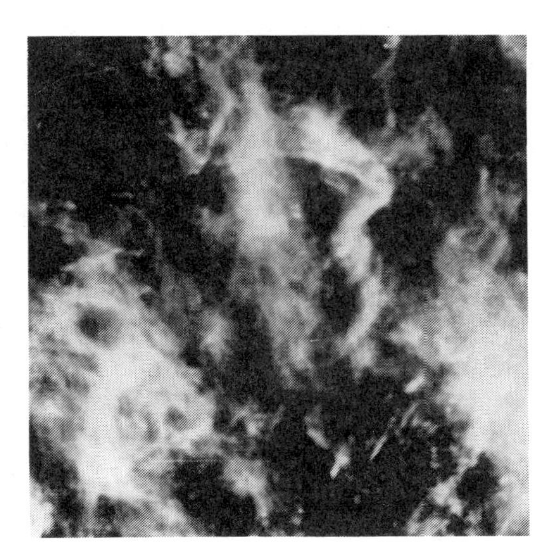

Figure 8.1: 100 μm map obtained with the *IRAS* satellite, showing the infrared cirrus; the map is 12.5° square. Reproduced courtesy of the Rutherford Appleton Laboratory and the Jet Propulsion Laboratory.

dust is not in equilibrium with the ambient radiation field.

Similarly, the discrepancy in the case of emission by interstellar grains is between the observed emission of grains and that predicted on the basis of the assumption that the emitting grains have an equilibrium temperature. The fact that theory and observation agree at the longest wavelengths suggests that the theory can not be completely wrong, and one way out of this dilemma is to suppose that there is a component of the grain population that can *not* be described as having an equilibrium temperature. We have already seen (Section 4.3.4) that such a situation can be provided by the stochastic heating process, in which small grains suffer extreme excursions in temperature as a result of the absorption of single photons. The excess emission seen at wavelengths $\lesssim 30\,\mu$m can be understood if individual grains are very small, with dimensions $\lesssim 10\,$Å and containing $\lesssim 100$ atoms; indeed such a small population of grains might be regarded as the very small particle limit of the MRN distribution, discussed in Section 6.4.5.

8.3 The nature of the small grains

8.3.1 Polycyclic aromatic hydrocarbons (PAHs)

We now consider the nature of these small grains. Note that one of the prime reasons for invoking very small grains is that the observations seem to imply that the infrared-emitting grains are not in equilibrium with the radiation field that heats them: an extreme, but transient, rise in temperature is required. The rise in temperature needed is such that small silicate or ice grains would immediately vaporize, ruling these materials out. We are then left, both on the grounds of abundance and on resilience to extreme excursions in temperature, with carbon as a major constituent of the small grain population. A cluster of some tens of C atoms is (like graphite) very likely to be planar, and we can expect such clusters to consist of a network of C atoms with H atoms around the periphery. Such clusters are known as *polycyclic aromatic hydrocarbons* (PAH) and their study has been of intense interest in recent years, not only in an astrophysical context but also in their chemical and physical properties. In particular, there has been a great deal of interest in these materials because of their possible relevance to emission features at infrared wavelengths. At the simplest level, these materials consist of combinations of a basic unit that resembles the benzene ring (C_6H_6); for example the PAH coronene ($C_{24}H_{12}$) consists of seven rings, while chrysene consists of four (see Fig. 8.6 below). From the above discussion such 'grains' can not be regarded as solid, and we discuss their nature and the nature of their infrared emission in this chapter.

8.4 Absorption and emission by PAH

The reason for the introduction of the PAH component of interstellar dust was to account for the excess infrared emission by interstellar grains. We have already determined the equilibrium temperature and consequent infrared emission of interstellar grains. Here we consider the mechanism whereby a PAH molecule emits infrared emission.

In general of course a molecule will have electronic energy levels, each of which is degenerate by virtue of the vibrational levels, each of which is, in turn, rendered degenerate by the rotational levels of the molecule. The energies involved in electronic, vibrational and rotational transitions are, to order of magnitude, in the ratios $[1 : (m_e/m_1)^{1/2} : (m_e/m_2)]$, where m_1, m_2 are the masses of the atoms or ions partaking in the vibration or rotation respectively. Typically, though not exclusively, electronic, vibrational and rotational transitions occur in the ultraviolet-optical, infrared

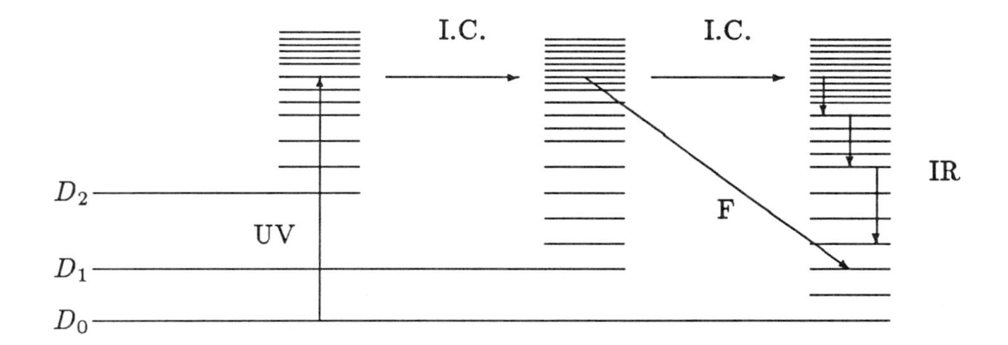

Figure 8.2: Internal conversion in a PAH molecule. Absorption of an ultraviolet (UV) photon is followed by internal conversion (I.C.) and by infrared emission (IR). Fluorescence (F) may also occur.

and millimetre-radio regions of the spectrum respectively.

8.4.1 Internal conversion

Superficially the production of infrared radiation by a PAH molecule proceeds in much the same way as the production of infrared emission by a 'normal' grain: in each case we have the absorption of an ultraviolet photon, followed by emission at infrared wavelengths. In the case of the PAH, however, the absorption and emission processes are constrained by the fact that the molecule has discrete energy levels and the dependence of the absorption efficiency on wavelength is (relatively) sharply peaked at wavelengths that are specific to the molecule; the same is of course true of the emitted infrared radiation.

We consider a PAH molecule in the ground electronic state D_0 (see Fig. 8.2). Absorption of an ultraviolet photon results in a transition to an excited electronic state D_2. Rather than (as might be expected) undergo de-excitation by cascading down the electronic levels back to the ground electronic state the molecule undergoes an isoenergetic internal conversion to a highly excited *vibrational* state of the ground electronic state (see Fig. 8.2). In such a transition there is no emission or absorption of energy, although a necessary condition for the internal conversion process to occur is that the transition probability for a spontaneous downward *electronic* transition be much less than that for the internal conversion transition. Once in the excited vibrational state of the ground electronic state the molecule cascades down the vibrational levels of the ground electronic state with the emission of infrared photons (infrared fluorescence). Alternatively the molecule may undergo internal conversion to an intermediate electronic state, followed by a transition to a lower vibrational level in the ground electronic state (fluorescence).

8.4.2 Temperature

The reason for the introduction of the PAH hypothesis was largely the discrepancy between the observed emission at wavelengths $\lesssim 30\,\mu$m and that predicted on the basis of the standard interstellar grain model: thermal 'spiking' of PAH-sized units was hypothesized to account for the difference. In this section we shall estimate the temperature to which a PAH molecule is raised following the absorption of an ultraviolet photon.

We should pause here and ask: What do we mean by the 'temperature' of a PAH molecule? For example if we fix our attention on a particular molecule in a gas at temperature T_{gas}, the gas–minus the molecule–can be regarded as a heat bath at temperature T_{gas}. However, the molecule will obviously not be in thermodynamic equilibrium with the gas: its energy will fluctuate erratically as it collides and exchanges energy with the other molecules in the gas. Clearly we can not speak of the 'temperature' of a molecule in the usual thermodynamic sense. We must therefore consider what precisely is meant by the 'temperature' of a PAH molecule.

Following the absorption of an ultraviolet photon the molecule ends up in highly excited vibrational states of the ground electronic state. In this state the molecule can be regarded as a system of independent harmonic oscillators of frequency ν_{i}, corresponding to the vibrational states of the molecule. We now show how this approach allows us to assign a 'temperature' T_{mol} to the molecule. We do this by determining the partition function for the vibrational system. For a simple harmonic oscillator with frequency ν in a heat bath at temperature T_{mol}, the partition function

$$Z = \frac{1}{1 - e^{-h\nu/kT_{\mathrm{mol}}}} \simeq \frac{kT_{\mathrm{mol}}}{h\nu},$$

where the approximation is valid for $h\nu \ll kT_{\mathrm{mol}}$. In the case of the PAH molecule the ultraviolet photon has provided the molecule with energy $U = h\nu_{\mathrm{UV}}$, resulting in a system of n independent harmonic oscillators, having frequency $\nu_1, \nu_2, \nu_3 \ldots \nu_{\mathrm{i}} \ldots \nu_{\mathrm{n}}$. For such a system the partition function may be written as

$$Z \simeq \frac{kT}{h\nu_1}\frac{kT}{h\nu_2}\frac{kT}{h\nu_3}\cdots\frac{kT}{h\nu_{\mathrm{i}}}\cdots\frac{kT}{h\nu_{\mathrm{n}}} = \frac{(kT)^n}{\prod_{i=1}^{n} h\nu_{\mathrm{i}}}. \tag{8.1}$$

The density of vibrational states $\rho(U)$ may be derived from the partition function (see Appendix C):

$$\rho(U) = \frac{U^{n-1}}{(n-1)!} \frac{1}{\prod_{i=1}^{n} h\nu_i}. \tag{8.2}$$

We now envisage a system, of internal energy U, which has the same physical components as the PAH molecule (n independent harmonic oscillators etc.) and which is in contact with a heat bath at temperature $T_{\rm mol}$. The temperature of this system is given by the thermodynamic relation

$$\frac{1}{kT_{\rm mol}} = \left(\frac{\partial \rho(U)}{\partial U}\right)_{\rm V,N}.$$

This is the temperature which we then assign to the PAH molecule.

Consider what happens when a PAH molecule absorbs a UV photon of energy $U = h\nu_{\rm UV}$ and enters an excited electronic state. As already discussed the molecule then undergoes internal conversion to highly excited vibrational states of the ground electronic state. In this state it is possible to regard the molecule as having internal energy U in contact with a heat bath at temperature T. Furthermore the molecule now consists of a system of independent harmonic oscillators, corresponding to the vibrational frequencies $\omega_i = 2\pi\nu_i$ of the molecule. A molecule containing N atoms has $3N - 6$ vibrational modes. This is precisely the situation described by the Einstein theory of specific heats, subsequently replaced, for the case of solids at least, by the Debye theory. We can therefore ascribe to the molecule a specific heat given by the Einstein approximation

$$C_{\rm V}(T) = k \sum_{i=1}^{3N-6} \frac{(\hbar\omega_i/kT)^2 \exp[\hbar\omega_i/kT]}{[\exp\{\hbar\omega_i/kT\} - 1]^2}; \tag{8.3}$$

the corresponding specific heat per mode is

$$c(T) = C_{\rm V}(T)/(3N-6). \tag{8.4}$$

We can now calculate the temperature $T_{\rm PAH}$ of a PAH molecule on absorbing an ultraviolet photon of energy $h\nu_{\rm UV}$; this is given by

$$h\nu_{\rm UV} = (3N-6)\int_0^{T_{\rm PAH}} c(T)dT. \tag{8.5}$$

The calculation of the temperature $T_{\rm PAH}$ requires a knowledge of the $(3N-6)$ modes of vibration of the PAH molecule and the determination of the corresponding $c(T)$ from Eqs. (8.3) and (8.4); the calculation is not trivial. For

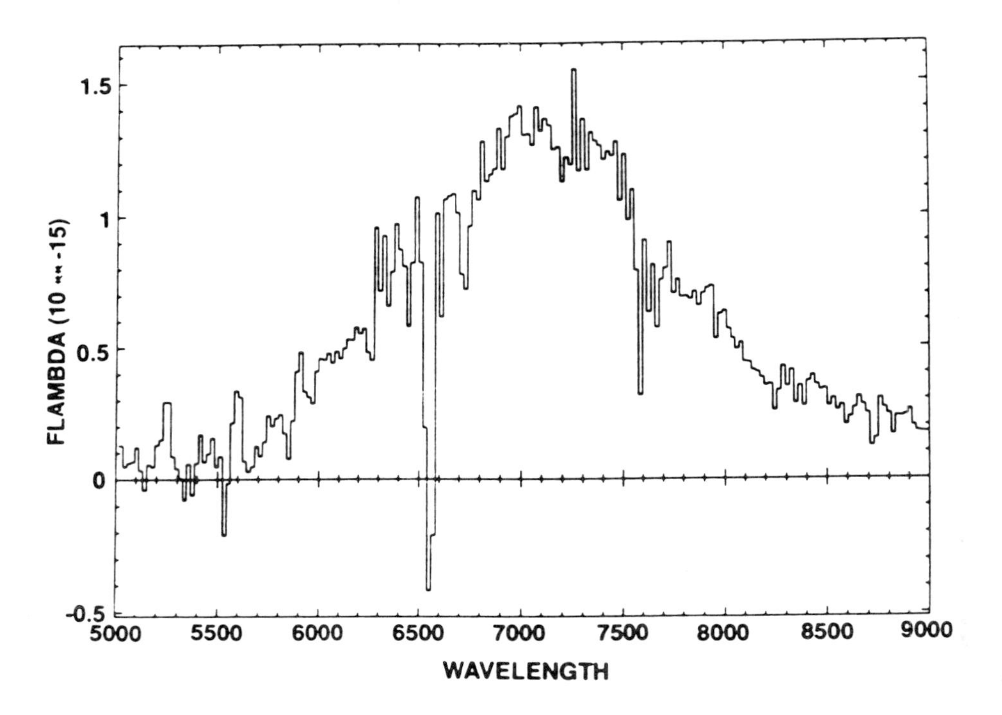

Figure 8.3: Extended red emission in the planetary nebula NGC 2327. Wavelength scale is in Å. After D. G. Furton & A. Witt, *Astrophysical Journal*, Vol. **386**, 587 (1992).

example, for a PAH molecule containing about 50 atoms, the peak temperature attained on absorbing a 10 eV photon is $\sim$ 1000 K. Alternatively, if $h\nu_{\rm UV}$ and $T_{\rm PAH}$ are known from observation, the number N of atoms in the PAH molecule can be estimated.

Extended red emission

A number of objects (reflection nebulae, HII regions) display a broad, shallow excess at the red (6600–7100Å) end of the spectrum and there is again much circumstantial evidence to associate this emission with dust particles in the vicinity of the illuminating star. The emission is particularly prominent for an object known as the Red Rectangle, a reflection nebula illuminated by a B9III star HD 44179; the excess emission in the planetary nebula[1] NGC 2327 is shown in Fig. 8.3. Such excess emission is even seen in the case of inter-

[1]NGC = New General Catalogue, a catalogue of nebulae and galaxies compiled at the end of the 19th century.

stellar dust and is often referred to as *extended red emission (ERE)*. Where ERE is associated with a particular star (as in the case of the Red Rectangle) the star is invariably of early spectral type, suggesting that ultraviolet photons are involved in the excitation of whatever is responsible for the emission; in the case of interstellar dust, excitation is by the background ultraviolet radiation. Fluorescence by PAH molecules has been suggested as the cause of ERE but a more likely identification is with hydrogenated amorphous carbon (HAC). This material consists of loosely aggregated clusters of carbon atoms, with hydrogen atoms occupying dangling bonds. Laboratory studies of photo-luminescence by HAC suggest that this is a strong candidate for the identification of ERE. Furthermore the degree to which the amorphous carbon is hydrogenated determines the band gap of the emitting material, and hence the wavelength of maximum emission; such variations in the spectral distribution of the ERE are indeed seen.

Infrared emission

It is perhaps the infrared emission of PAH that has attracted most attention because of the presence, in the infrared spectra of a variety of objects, of a number of emission features that had eluded identification. Indeed for many years these features were referred to as the 'unidentified infrared features' but, following the surge of interest in PAH and their infrared emission, are now sometimes called the 'overidentified infrared features'. Fig. 8.4 shows a typical selection of the unidentified features, in the infrared spectrum of the planetary nebula NGC 7027, although they are seen in a wide variety of astronomical objects.

Just as any material (such as olivine) with a Si–O bond is expected to give rise to a feature at 9.7 μm, so any material which contains the C–H bond, including PAH and HAC, will give rise to an emission feature at 3.28 μm. The stretching of the C–H bond gives rise to the feature at 3.28 μm, while the out-of-plane bend of the bond gives rise to a feature at 11.25 μm. Table 8.1 lists some of the 'unidentified' features seen in the infrared spectra of astronomical objects (cf. Fig. 8.4), together with the suggested identification on the PAH interpretation.

The diffuse features

The presence of diffuse absorption features in the spectra of reddened stars has been mentioned in Chapter 1. We have already noted that the carrier(s) of these features seem to be well-mixed with (or may even be part of) the interstellar dust population. Although some of the features are broad, in-

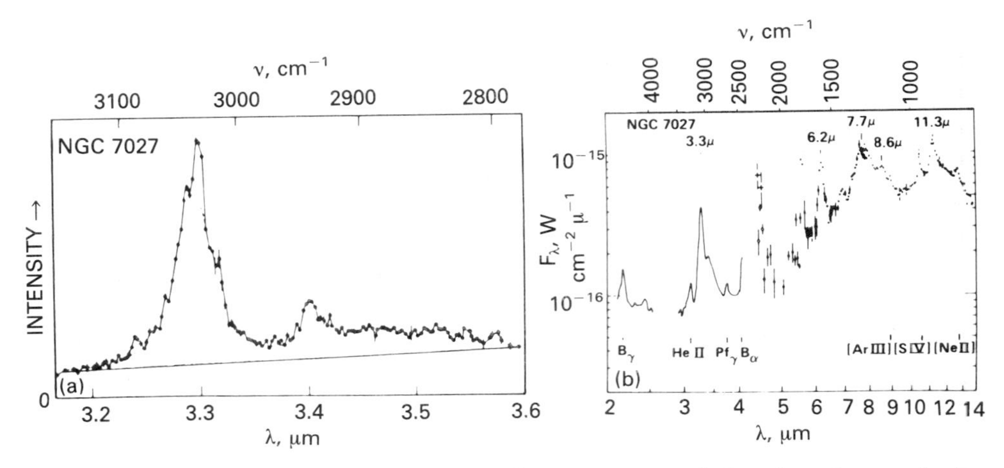

Figure 8.4: The infrared features in the spectrum of the planetary nebula NGC 7027. After L. J. Allamandola et al., *Astrophysical Journal Supplement*, Vol. **71**, 733 (1989).

dicating an origin in the solid rather than the gas phase, the narrowness of others indicates a molecular, rather than solid, origin. The advent of the PAH hypothesis led several workers to consider the possibility that these molecules might be responsible. In testing this hypothesis we are faced with the usual difficulty of reproducing likely cosmic PAH species in the laboratory (see Section 8.7 below); nonetheless there have been claims that singly-ionized PAH molecules could account for some of the diffuse bands.

8.5 Physical properties of PAH

In Chapter 2 we described the physical properties of dust particles, including their temperature and electric charge. In this section we do the same for the PAH.

8.5.1 Dimensions

The dimensions of a PAH molecule will obviously depend on its exact nature: a PAH like coronene is approximately circular whereas chrysene ($C_{18}H_{12}$) is more linear (see Fig. 8.6). The linear PAHs are generally less stable so that the circular PAHs are likely to predominate if they do indeed exist in interstellar space. Since the C–C bond is 1.4 Å long simple geometry suggests that each hexagonal unit has area 5 Å^2. A near-circular PAH, having n (n odd) hexagonal units along its greatest dimension (e.g. coronene has $n = 3$;

Table 8.1: 'Unidentified' infrared emission features

$\lambda(\mu m)$	Assignment
11.25 μm	C–H out-of-plane bend
8.6 μm	C–H in-plane bend
7.7 μm	C–C stretch
6.2 μm	C–C stretch
3.4 μm	C–H stretch
3.28 μm	C–H stretch
6600Å (ERE)	HAC

see Fig. 8.6), has a total of $3n - 3$ hexagonal units. Furthermore the PAH has $n_C \simeq 3(3n - 3)$ carbon atoms, provided $n_C \gg 1$ [not $n_C = 6(3n - 3)$ otherwise we would be counting each C atom twice]. Thus the total area of such a PAH molecule is $\simeq 15(n - 1)$ Å^2, or $\sim 2n_C$ Å^2. The radius of a PAH molecule containing n_C ($\gg 1$) carbon atoms is therefore $\sim 0.7 n_C^{1/2}$ Å.

The cross-section of a PAH molecule for the absorption of ultraviolet and visible radiation is generally estimated by determining the cross-section per carbon atom appropriate to graphite *particles* (to which Mie theory may be applied), and assuming that extrapolating this to smaller sizes gives the cross-section per carbon for a PAH molecule; the possible effect of the peripheral H atoms is neglected in this approximation. From the discussion of Section 3.4.2 the absorption cross-section of a graphite sphere of radius a is $Q_{\rm abs}\pi a^2$ so that the cross-section per carbon atom is

$$\sigma_{\rm C} \simeq \frac{3 m_{\rm C} Q_{\rm abs}}{4 a \rho},$$

where m_C is the mass of a carbon atom. The absorption cross-section for a PAH molecule containing n_C carbon atoms is therefore

$$\sigma_{\rm PAH} \simeq n_{\rm C} \frac{3 m_{\rm C} Q_{\rm abs}}{4 a \rho}.$$

This is the relevant cross-section for the absorption of ultraviolet radiation when we discuss the temperature of a PAH molecule.

The cross-sections σ_i for the *ionization* of PAH molecules, relevant to the discussion in the following subsection, have been extensively measured in the laboratory. The results of these measurements suggest that, to a reasonable approximation, the ionization cross-section of a PAH molecule scales simply as the number of carbon atoms it contains, and this conclusion is borne out by theoretical considerations. Numerically

$$\sigma_{\rm i} \simeq 10^{-22} n_{\rm C} \ \ {\rm m}^2.$$

Again this assumes that the effect of peripheral H atoms is negligible, and that the photoionization yield Y is close to unity. The latter assumption should be reasonable for molecules of this size: it is only in bulk solids that one expects the ejected photoelectrons to be stopped before they reach the surface, thus rendering $Y < 1$.

Electric charge

As in the case of a conducting grain near a source of ultraviolet radiation, a PAH molecule will be positively charged. However, unlike the case of a grain the photoelectric effect will not be the relevant process: instead a PAH molecule will be photoionized. The first ionization potential of PAH molecules is expected to be typically $\sim 5 \rightarrow 10\,\text{eV}$; the second ionization potential is probably very much higher than this so in this case a PAH molecule is generally expected to carry unit positive charge. The electron is expelled from the highest occupied π orbital.

PAH molecules can also acquire negative charge by the addition of electrons. However, unlike the situation for spherical grains (discussed in Section 4.2) we do not expect PAH molecules to acquire electric charge by the sticking and removing of electrons to the 'surface'. Electrons will be added to the lowest unoccupied π orbital. The distribution of electrons *within* the PAH molecule will result in the molecule being polarized, mainly as a result of the π electrons. The contribution of the π electrons to the polarizability of the PAH molecule varies approximately as the cube of the dimensions of the molecule.

8.6 The relation between PAH and larger grains

PAH molecules may collide and aggregate into larger clusters provided that the binding energy of the PAH–PAH pair exceeds the kinetic energy of the individual PAH molecules in their centre of mass system. This condition requires that the colliding PAH molecules be electrically neutral, otherwise the centre of mass energy needed to overcome the Coulomb barrier will exceed the binding energy of the resultant pair. The newly-formed PAH–PAH system then relaxes radiatively. Quantum mechanical calculations of the binding of identical PAH molecules show that certain configurations are

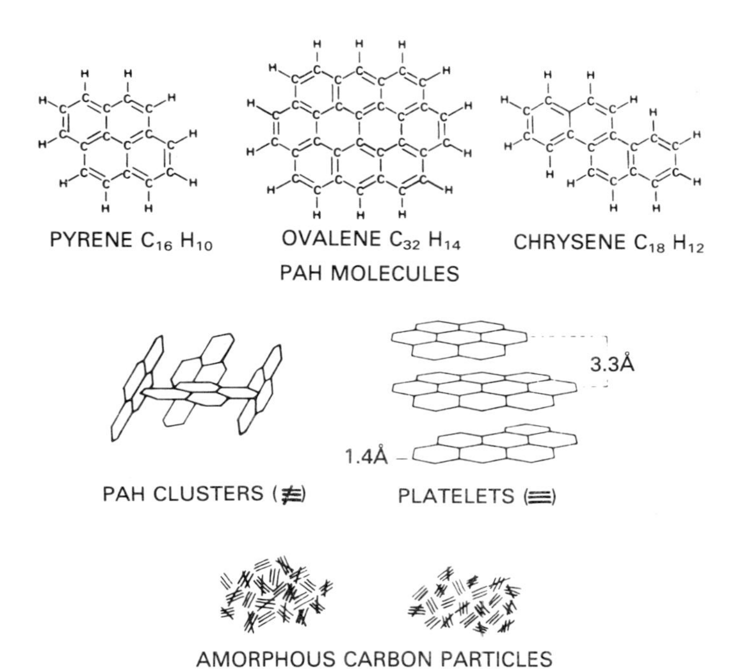

Figure 8.5: The relationship between PAH molecules and amorphous carbon. After L. J. Allamandola et al., *Astrophysical Journal Supplement*, Vol. **71**, 733 (1989).

preferred; these include the end-to-end configuration (— —), face-to-face (||), angular (∧), and tee (⊥). The tee and face-to-face geometries seem to be one of the most stable of these configurations, with the result that PAHs are expected to aggregate in these forms. Such aggregates will then be bound together by chemical bonds formed by bridging groups, or by van der Waals forces, and the resulting material will be amorphous carbon (see Fig. 8.5). Indeed there is a very close relationship between this material and the HAC discussed above.

8.7 Problems with the PAH hypothesis

The PAH hypothesis was put forward in order to account for discrepancies between the observed and calculated infrared emission of interstellar dust, and the anomalous behaviour of the temperature of the dust in reflection nebulae. The fact that PAH molecules have emission features in the infrared that coincide–even if only approximately–with the hitherto unidentified infrared emission bands was a bonus. Nevertheless a significant number of workers remain unconvinced by the PAH hypothesis and we examine some of their misgivings in this section.

8.7.1 The infrared emission bands

The infrared emission spectrum of the 'bar' in the Orion nebula is illustrated in Fig. 8.6, in which it is compared with the emission of a number of PAH molecules. As with many of these comparisons the agreement is good, but not precise. Indeed the unidentified infrared emission bands in the spectra of astronomical sources have never been satisfactorily reproduced using laboratory data. However we should recall, as always, that it is extremely difficult to produce samples in terrestrial laboratories that bear any resemblance to materials present in astronomical environments.

8.7.2 Ultraviolet features

As already discussed the key process in exciting PAH emission in the infrared is the initial absorption of an ultraviolet photon followed by an isoenergetic transition from an excited electronic state to high vibrational states of the ground electronic state. The assumption is that the isoenergetic transition is more probable than ultraviolet fluorescence. However there is some experimental evidence to suggest that this may not be the case and that ultraviolet fluorescence will in fact occur. Therefore if PAH molecules are indeed a substantial component of the interstellar medium we may expect, as a consequence of the above discussion, that the interstellar extinction curve shows evidence of absorption features in the ultraviolet corresponding to excitation from the ground electronic state, together with the infrared features that arise as the molecule cascades down to the ground vibrational state. The presence of strong features in the wavelength range 2000–2500 Å is characteristic of these materials although, as we have seen, there are no such features in the interstellar extinction curve. However, where it has been possible to determine the extinction curve of newly condensed *circumstellar* carbon dust (e.g. in the winds of erupting novae), there seems to be evidence for a feature centred at 2500 Å. While there are difficulties in determining the extinction law for circumstellar dust [e.g. neither the foreground interstellar extinction nor the intrinsic (unreddened) flux distribution of the star may be sufficiently well known] these results may hint at the presence of PAH-like nucleation sites. In general, however, the fact that the expected features in the 2000-2500 Å range are not observed is commonly put forward

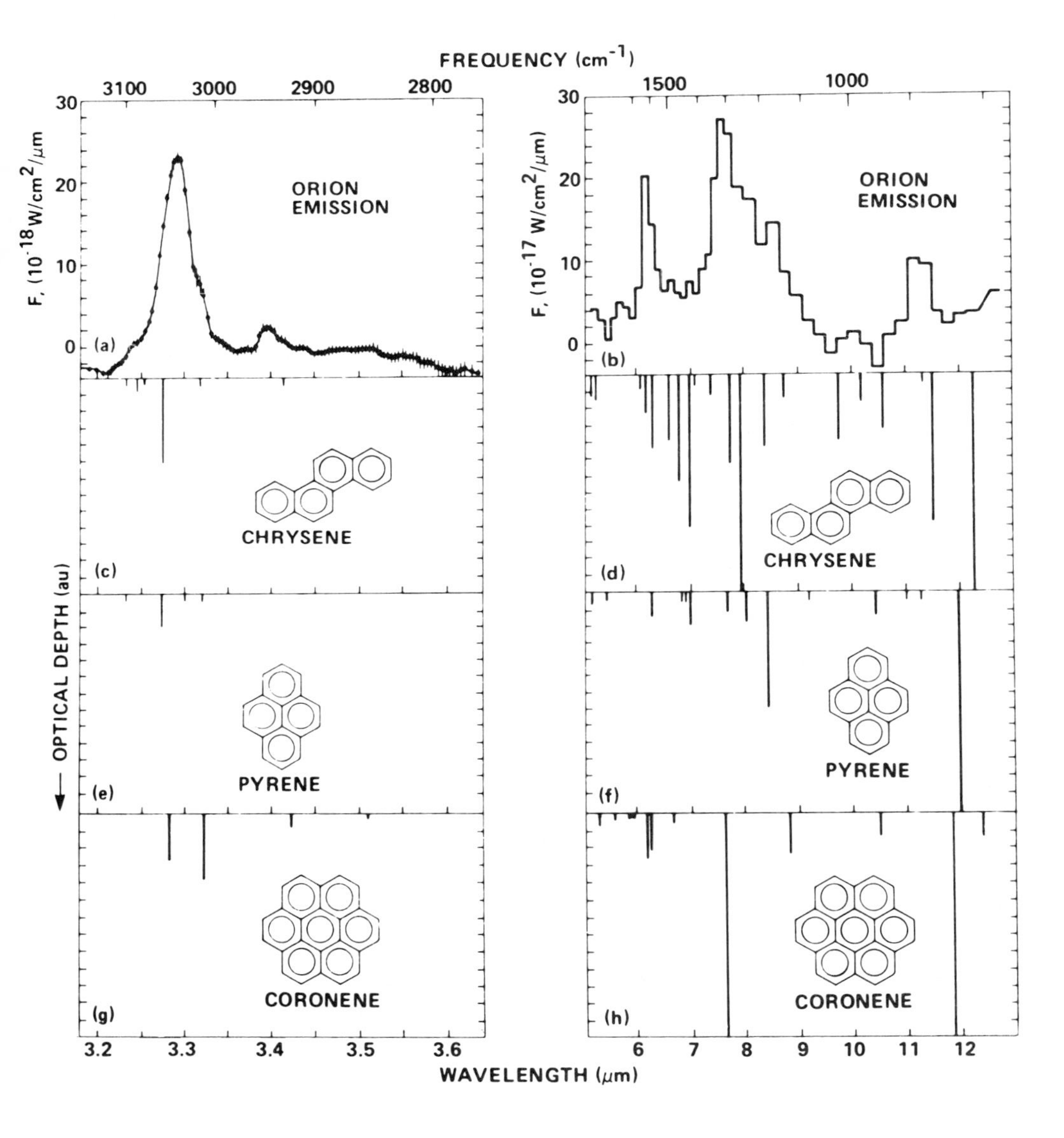

Figure 8.6: The infrared spectrum of the Orion bar compared with the laboratory infrared spectrum of the PAH molecules. After L. J. Allamandola et al., *Astrophysical Journal Supplement*, Vol. **71**, 733 (1989).

as a serious objection to the PAH hypothesis. Proponents of the PAH hypothesis counter that the mix of PAH molecules likely to exist in interstellar space would result in the 'smearing out' of the individual ultraviolet features and the resultant absorption would be broad, shallow and undetectable in the total extinction curve.

8.7.3 Laboratory and cosmic PAHs

In discussing the PAH hypothesis we are constrained to comparing astronomical data (in this case mainly in the form of infrared and ultraviolet spectroscopy) with laboratory data and we should, as always, take care that we are comparing like with like: how sure are we that the PAH material studied in the laboratory resembles that possibly present in interstellar and circumstellar space?

As is widely recognized, it is never easy–sometimes impossible–to simulate cosmic materials in the laboratory. In the case of PAHs the following points are almost certainly relevant. In the laboratory the number of atoms per PAH is $\lesssim 40$ whereas in astronomical environments the number of atoms per PAH is $\sim 50 \rightarrow 100$. Furthermore cosmic PAHs are isolated and almost certainly ionized, whereas those investigated in the laboratory are neutral and generally in solid form.

Laboratory measurements generally measure the transmission of candidate cosmic materials, which if anything measures the *extinction* rather than the *absorption* properties of the material; in the case of PAHs of course–in which laboratory analogues are compared with the infrared emission features–such a comparison seems inappropriate. In addition it is common to mount the sample to be measured in or on a spectroscopically neutral substrate (e.g. KBr). While the substrate material has neutral transmission in the wavelength range of interest, there has to be some concern that the incorporation of the sample into the substrate may completely alter the physical characteristics of the material to be measured.

8.8 Large carbon molecules

Laboratory measurements of 'soots' produced in the process of carbon combustion show that certain carbon molecules are produced which are extremely stable; in particular the molecule C_{60} seems especially favoured. The structure of this molecule is believed to resemble a soccer ball, with carbon atoms at the vertices of the surface pentagons and hexagons (see Fig. 8.7); other large carbon molecules (such as C_{72}) are believed to have similar structure. The molecule is sometimes referred as 'soccerene' but more

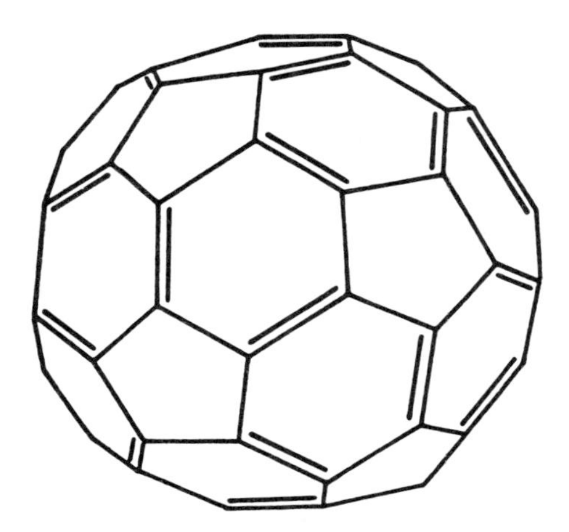

Figure 8.7: The fullerene molecule C_{60}. After H. Kroto, *Science*, Vol. **242**, 1139. ©1988 American Association for the Advancement of Science.

usually as 'buckminsterfullerene'. The physical properties of C_{60} have been the subject of intense study recently, from the point of view of its possible technological applications, and astrophysics has inevitably benefitted from such study.

C_{60} opens up a number of fascinating possibilities. For example, the ^{22}Ne isotope is found in certain components of carbonaceous meteorites. There are strong grounds for supposing that this isotope arises from the β-decay of ^{22}Na, which has a radioactive half-life of 2.6 yr. The short lifetime of the ^{22}Na isotope must therefore imply that it is produced and 'frozen' in a dust grain in a very short time ($\simeq$ a year); one possibility is that it is trapped inside a fullerene molecule which is subsequently subsumed in a dust grain, for example in a nova explosion (see Chapter 7).

Reading

The following provide excellent reviews of the PAH hypothesis; the first two are sympathetic to the PAH hypothesis while the third includes both pros and cons:

[D] L. J. Allamandola, A. G. G. M. Tielens & J. R. Barker, *Astrophysical Journal Supplement*, Vol. **71**, 733 (1989).

[D] J. L. Puget & A. Léger, *Annual Review of Astronomy and Astrophysics*, Vol. **27**, 161 (1989).

[D] *Interstellar Dust*, Proceedings of International Astronomical Union Symposium 135, Eds L. J. Allamandola & A. G. G. M. Tielens, Kluwer Academic Publishers (1989).

The following paper gives a very full discussion of the physics of PAH molecules:

[D] A. Omont, *Astronomy & Astrophysics*, Vol. **164**, 159 (1986).

9

Extragalactic dust

9.1 Introduction

As discussed in previous chapters, the evidence for dust grains in the Galaxy, between and around stars, is beyond question. We briefly describe in this chapter the evidence for dust beyond the Galaxy, both distributed uniformly between galaxies (the intergalactic equivalent of interstellar dust) and within the galaxies themselves. Since the evidence for the latter is easily obtained, we consider the observation and properties of dust within galaxies first, before turning to the possibility that there might exist a diffuse distribution of intergalactic dust. However, just as we did in Chapter 1, we first review briefly the general background.

9.2 Galaxies

9.2.1 The Hubble sequence

Since the nature of the galaxies was elucidated early in the 20 th century there have been several schemes for classifying them according to their morphology, stellar content etc., but the essential features of all these schemes are based on E. P. Hubble's classification of the 1930's.

Elliptical galaxies are, as the name implies, elliptical in appearance and are classified as En, where n is a number that describes the ellipticity of the galaxy as follows. If the major axis of the galaxy image is designated by a and the minor axis by b then

$$n = 10\frac{(a-b)}{a}$$

The value of n is, of course, determined not only by the intrinsic geometry of the galaxy but also by the orientation of the galaxy relative to the observer. In general, elliptical galaxies tend to contain Population II stars and, until the *IRAS* mission, were generally believed to be devoid of dust, although notable exceptions to this rule were known.

Spiral galaxies have a nucleus from which two or more spiral arms emanate. They are classified according to the 'tightness' of the spiral arm and the prominence of the nucleus. They are further subdivided according as to whether they are normal spirals (S) or barred spirals (SB), the spiral arms in the latter type originating at the ends of a 'bar' rather than in the nucleus itself. A normal spiral having a prominent nucleus and tightly wound spiral arms is designated Sa, while a normal spiral with a small nucleus and loosely wound spiral arms is designated Sc; a type Sb spiral has a moderately prominent nucleus and spiral arms of moderate tightness. A similar scheme applies to barred spirals; thus for example a barred spiral with a prominent nucleus and tightly wound spiral arms is classified as SBa. As a rule, the nucleus of a spiral consists of Population II stars and the nucleus itself is gas and dust free; furthermore a spiral galaxies possess a 'halo' of Population II stars. The spiral arms are gas and dust rich and contain high luminosity Population I stars.

Irregular galaxies, as their description implies, defy classification in either of the above categories. They are often chaotic in appearance and consist of Population I stars, although Population II stars are also present. As the Population I content might imply, they also tend to contain a substantial amount of free gas.

The dimensions of galaxies are in the range 10–50 kpc, and galaxy masses are typically $\sim 10^{11} M_\odot$ (in other words, a typical galaxy contains some 10^{11} stars). Fig. 9.1 illustrates the various types of galaxy described above.

In general, the flux distribution of these galaxies, at near infrared wavelengths and shortwards, is simply the sum of the flux distributions of all the stars etc. that the galaxies contain. Notwithstanding the different stellar populations, the flux distributions in the optical and near infrared are essentially stellar, i.e. thermal, in nature. It is at longer wavelengths, where the stellar contribution becomes negligible, that emission by the dust content of the galaxy becomes apparent.

9.2.2 Other extragalactic objects

As well as the galaxies explicitly described by the Hubble scheme, there are other types of extragalactic objects that do not fit the above categories.

Figure 9.1: Group of galaxies, including an elliptical (upper left), a spiral (upper centre) and a barred spiral (lower right). Note the heavy obscuration in the spiral near the centre of the field. Photograph from the Hale Observatories.

Quasars resemble stars on photographic plates but emission lines in their spectra are grossly shifted to the red. This is generally interpreted in terms of the expansion of the Universe (see below), in which the quasars are assumed to partake. Quasars, and the closely related *blazars*, are believed to be very early stages in the evolution of galaxies. Unlike normal galaxies, the flux distributions of quasars and blazars are distinctly *non-thermal*, and may frequently be represented by a power-law continuum $S_\nu \propto \nu^{-\alpha}$ with (in the case of quasars) emission lines superimposed. Such a continuum is generally produced by a process such as synchrotron radiation, or inverse compton scattering.

Seyfert galaxies are generally spiral in form but are distinguished by having exceptionally bright, quasar-like, nuclei. The nuclei emit non-thermal radiation although in some Seyferts, the infrared emission is thermal. Seyfert galaxies, quasars etc. are generally grouped together under the general heading of *Active Galactic Nuclei*, or AGN.

Starburst galaxies are galaxies (usually spiral or irregular) which seem to be undergoing an outburst of intense star-forming activity. In some cases this activity may be triggered by the interaction of two galaxies.

9.2.3 Clustering of galaxies

Galaxies generally aggregate into clusters and our own Galaxy is no exception: it belongs to a small group–the *Local Group*–of galaxies which is about 1 Mpc in diameter and contains about 25 members. From our point of view the most important members of the Local Group are the Large and Small Magellanic Clouds–irregular galaxies that are satellites of our own–and the large Sb spiral galaxy M31.

Other clusters are very much larger than this; for example, the Coma cluster, distant some 90 Mpc, contains about 800 members and is some 6 Mpc in diameter. There is also some evidence that *super*-clustering may occur–i.e. clusters may form clusters of clusters. Furthermore, there is mounting suspicion that, far from being uniformly distributed, galaxies and clusters may be distributed in the form of 'sheets'; however, none of this need concern us here.

9.3 Some cosmological preamble

The discussion of real extragalactic dust–as opposed to the dust within galaxies–imposes a complication that is not present in the interstellar case, namely the fact that the Universe is expanding. Accordingly we provide here a brief summary of the relevant ideas so that the results of the present chapter can be viewed in context. No attempt is made to be either rigorous or complete: the interested reader is referred to the bibliography.

9.3.1 The expansion of the Universe

Once we get significantly beyond the Local Group we find that galaxies and clusters of galaxies are receding from us; this is the *expansion of the Universe* discovered by V. M. Slipher and E. Hubble in the 1920's. The expansion is such that, for recessional velocities small by comparison with the velocity of light, the recessional velocity V of a galaxy at distance D is given by *Hubble's Law*:

$$V = HD, \tag{9.1}$$

where H is Hubble's constant[1]. There is still considerable argument as to the value of H; in this chapter we shall assume a value $H = 75\,\mathrm{km\,s^{-1}\,Mpc^{-1}}$.

[1]More correctly, the Hubble parameter; Hubble's constant is not, in fact, constant.

9.3.2 The redshift

We have already encountered the idea that the Universe is in a state of expansion and that, provided the galaxies concerned are not too distant, the expansion can be described by Hubble's law [Eq. (9.1)]. Suppose we measure a spectral line, whose laboratory wavelength is $\lambda_{\rm em}$, in the spectrum of a distant galaxy. Its wavelength in the spectrum of the galaxy is measured to be $\lambda_{\rm obs}$, which is always greater than $\lambda_{\rm em}$; in other words the spectral line is shifted to the red end of the spectrum. The *redshift* z is defined as

$$1 + z = \frac{\lambda_{\rm obs}}{\lambda_{\rm em}}, \tag{9.2}$$

or

$$z = \frac{\delta\lambda}{\lambda_{\rm em}} = \frac{\lambda_{\rm obs} - \lambda_{\rm em}}{\lambda_{\rm em}}, \tag{9.3}$$

where $\delta\lambda$ is the change in wavelength caused by the expansion of the Universe[2]. At the time of writing the redshift record for any object is held by the quasar PC 1247+3406, which has a redshift $z = 4.897$.[3] The shift $\delta\lambda$ was originally interpreted in terms of the Doppler effect, for which

$$\frac{\delta\lambda}{\lambda} \simeq \frac{V}{c}$$

for $V \ll c$.

If we take Hubble's Law at face value, we might conclude that there is a certain distance, given by c/H, at which galaxies would be receding at the speed of light. However, we should not take Hubble's Law too literally; what we can say is that Hubble's law suggests a scale for the Universe of $\sim c/H \simeq 4$ Gpc. Another implication is that the expansion began a time $H^{-1} \sim 13 \times 10^9$ years ago. Again we should not take this conclusion too literally. However, detailed calculation does indeed show that the age of the Universe is of the order of, but cannot exceed, H^{-1}.

It is reasonable to ask at this point whether the expansion of the Universe will continue indefinitely, or whether the expansion will eventually reverse into a collapse. The answer to this question depends on the mean density of the Universe, as measured in our vicinity. If the mean density exceeds a certain critical density $\rho_{\rm C} \simeq 10^{-26}\,{\rm kg\,m^{-3}}$, then the present expansion will indeed come to an end and the Universe will collapse; however, if the mean

[2]Now that astronomy is carried out in most of the accessible electromagnetic spectrum, the term redshift is generalized to mean a shift to *longer* (not just red) wavelengths. This is to avoid ambiguities in (for example) the infrared and radio bands

[3]The 'surface of last scattering', from which the cosmic microwave background originates (see Section 9.3.3), is at a redshift of about 1500.

density is less than ρ_C the expansion will continue indefinitely. Unfortunately the mean density of the Universe is very difficult to determine since we are constrained to measure the density in the form of luminous matter (galaxies etc.), but there are indications that a substantial amount of non-luminous matter may be present (the so-called dark matter problem).

The light year as a unit of distance was mentioned in Chapter 1. In the cosmological context however we have to be careful when we speak of distance. As we look out on the Universe we look back in time: by definition we see an object whose 'distance' is X light years as it was X years ago. For objects within the Galaxy–and even for the nearer galaxies–this concept is perfectly acceptable and agrees with whatever other concept of distance we care to use. In an expanding Universe, however, in which we might be looking back over a substantial fraction of the age of the Universe, such a measure of distance will not suffice and indeed, we can define distance in a number of different ways. However we need not become involved with the details of the various definitions of distance here. Fortunately, whichever measure of distance we choose to use has the property that it increases monotonically with redshift, defined by Eq. (9.2); indeed, for small enough redshifts ($z \ll 1$), $D \simeq cz/H_0$, irrespective of cosmological model and definition of distance. Therefore whenever we wish to talk about the 'distance-dependence' of any quantity we will generally refer to its dependence on redshift z.

9.3.3 The '3K' background

As the Universe expands it thins out: the mean density of any given portion of the Universe decreases with time. This might suggest that the expansion began from a state of extremely (possibly infinitely) high density and temperature a time $\sim H^{-1} \simeq 13 \times 10^9$ years ago; the high density, high temperature state of the Universe gives rise to a phenomenon which is relevant insofar as a study of dust in the Universe is concerned.

We have already mentioned on several occasions the fact that hydrogen comprises about 73 per cent of cosmic matter by mass, with helium comprising about 25 per cent. During the first 200 seconds or so of cosmic time the Universe consisted of protons, neutrons and electrons (as well as photons and neutrinos), and was dense enough and hot enough for the thermonuclear fusion of hydrogen to helium: the process that currently goes on in the interiors of stars like the Sun was occurring on a Universal scale. However, the expansion (and consequent cooling and thinning out) of the Universe prevented the synthesis of heavier elements. As a consequence, it is estimated that about 25 per cent of cosmic matter, by mass, should be in the form of helium and this is indeed the case. The inference is that the cosmic helium we now see (but little else) was synthesized during the first few moments of the Universe.

Subsequently the Universe cooled as it expanded, but while the temperature remained above $\sim$ 4000 K, hydrogen was ionized and the Universe was opaque as a result of Thomson scattering of radiation by free electrons. As soon as the Universe cooled below 4000 K, protons and electrons 'recombined'[4] and the Universe became transparent. Since the velocity of light is finite, the further out into the Universe we look the further back we go in time. We can in principle therefore look back to the point in the Universe's history–less that one million years after its origin–when transparency occurred. We should see a surface that covers the entire sky and which emits blackbody radiation at about 4000 K; however, the expansion of the Universe has redshifted the radiation from this surface so that it now appears to us as a surface at $\sim$ 2.7 K; this is the 'cosmic microwave background', already mentioned several times. Indeed the temperature of this surface has been measured very accurately by the *COBE* satellite to be $T_{\rm cos} = 2.735\,{\rm K}$.

9.4 Dust in galaxies

We can approach the search for dust within galaxies in three ways. First, we can take the simple approach and simply look at galaxies to see if there is any evidence of obscuring material. As we can see in photographs of the Milky Way, there is direct evidence for dust in the form of dark lanes and rifts (recall Figs. 1.1 and 1.9). This would seem to be an approach we could use in the case of galaxies and indeed, there is ample photographic evidence for dust in galaxies. Examples are given in Figs 9.1 and 9.2; in the latter in particular a dark lane of dust is clearly seen silhouetted against the stellar light of the galaxy. Second, we can look at individual stars in a galaxy, make allowance for the effect of interstellar dust in our own Galaxy and see if there is any evidence of wavelength-dependent extinction. Obviously this method requires that individual stars in a galaxy can be resolved and, in practice, this means galaxies in the Local Group–mainly the Large and Small Magellanic Clouds. Third, if a galaxy contains any dust at all, we might expect that the dust is heated–either by the stars in the galaxy or by some non-stellar 'power-house' lurking in the galaxy's nucleus. The third method will of course require observations at infrared wavelengths, and this has been particularly fruitful in recent years.

[4]It may seem strange to speak of *re*-combination of protons and electrons when they weren't 'combined' in the first place.

Figure 9.2: A galaxy with a dust lane: NGC 5128. The supernova 1986G is arrowed. Photograph by David Malin. © Anglo-Australian Telescope Board.

9.4.1 Extinction in other galaxies

For galaxies sufficiently close to us that individual stars can be resolved, it may be possible to determine the extinction law for the dust population of the galaxy concerned. In practice this requires not only that individual stars be resolved but also that they be sufficiently bright over a wide range of wavelengths so that the wavelength-dependence of the extinction law can be determined. A further requirement is that the stars be bright enough optically that their spectral types can be determined from their spectra. The procedure then is to observe (i) a reddened star in the chosen galaxy and (ii) an unreddened star of the same spectral type. As already noted, this means that studies of this kind have to be confined to galaxies, such as the Magellanic Clouds, in the Local Group and to highly luminous stars of early spectral type. Even so such work that has been done may point to possible differences between the interstellar dust populations of these galaxies and our own.

The study of the extinction laws for dust in the Magellanic Clouds is es-

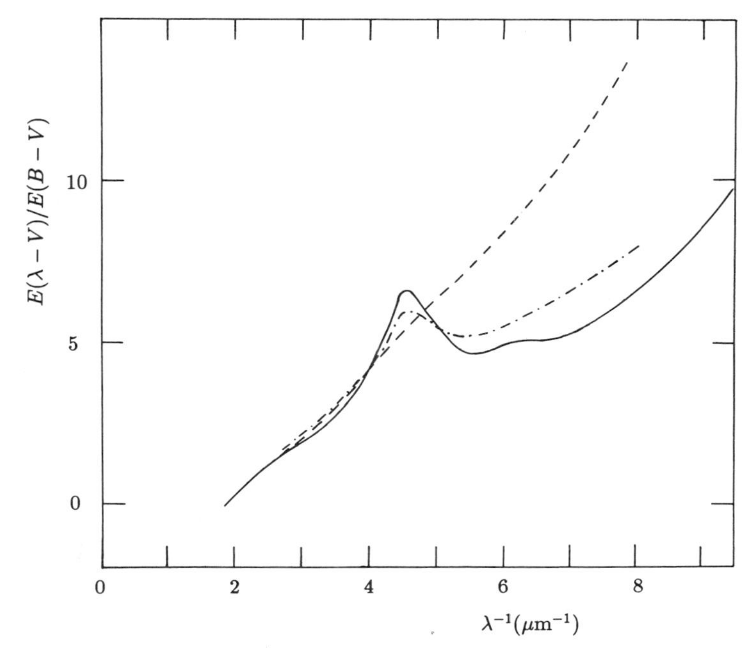

Figure 9.3: Extinction laws in the Galaxy (full curve), the Large Magellanic Cloud (dash-dot curve) and the Small Magellanic Cloud (dashed curve). Adapted from E. L. Fitzpatrick, *Interstellar Dust*, IAU Symposium 135, p. 37 (1989).

pecially intriguing as the interstellar gas in both these galaxies have elemental abundances that may be substantially different from those appropriate to our Galaxy. For example the abundance of oxygen in Large Magellanic Cloud (LMC) HII regions is a factor of two times *less* than in Galactic HII regions, whereas in Small Magellanic Cloud (SMC) HII regions it is a factor of five lower. Such differences in the chemical abundances of the interstellar gas in other galaxies may hint at possible differences in the interstellar dust population (cf. Table 6.2), and hence in their respective extinction laws.

Early work on the extinction law in the LMC suggested that there are indeed distinct differences between it and the Galactic extinction curve, in that the '2175' extinction bump is weaker and the rise in extinction to the far ultraviolet is much steeper for LMC dust than for dust in the Galaxy. However, it transpired that the LMC stars for which these extinction curves were obtained were concentrated in one region of the LMC and the extinction law in other regions did not differ so substantially from that in our own Galaxy. In the case of the SMC it seems that the extinction law lacks the '2175' feature seen in the Galactic extinction law and indeed, the extinction

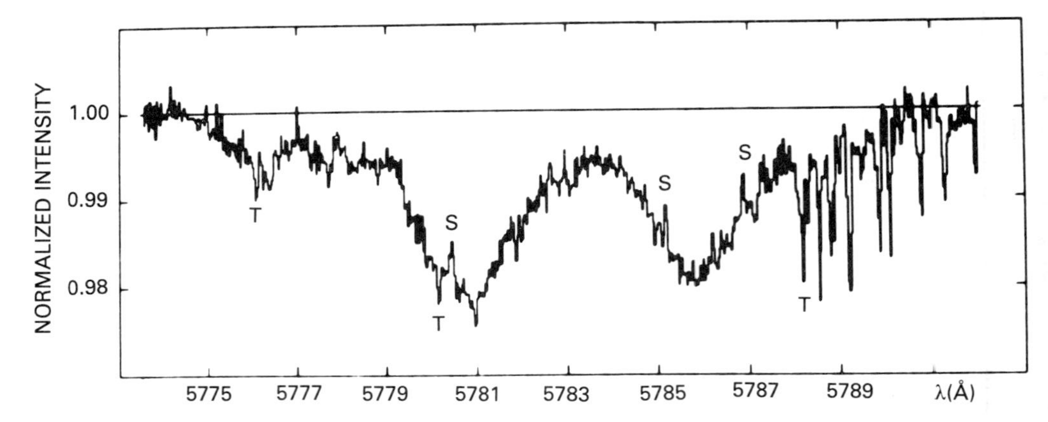

Figure 9.4: The 5780.4 Å diffuse band arising in interstellar dust in our own Galaxy and (shifted to longer wavelengths) in the LMC. An 'S' denotes an instrumental spot, a 'T' an absorption line arising in the Earth's atmosphere. After G. Vladillo et al., ***Astronomy & Astrophysics***, Vol. **182**, L59 (1987).

seems to follow a simple λ^{-1} law throughout the visible and ultraviolet. However in this case also the results have to be approached with caution as extinction laws are available for only a handful of SMC stars. The extinction laws are illustrated schematically in Fig. 9.3.

There have been a number of searches for the interstellar diffuse bands in the dust populations of other Local Group galaxies. The above requirements for investigating the extinction laws in other galaxies can be relaxed somewhat in the case of the diffuse bands because it is required only that the background star be *optically* bright. The fact that the same diffuse bands may be present in the dust in both the host galaxy in our own poses no difficulty. If there are indeed diffuse bands in the interstellar dust of other galaxies they will share the motion of their host galaxy relative to our own; the diffuse bands will therefore be doppler shifted and readily distinguishable from 'local' diffuse bands arising in our own Galaxy.

The search for diffuse features in the dust of other galaxies has involved observing the spectra of extragalactic supernovae and, as expected, the supernova 1987A in the LMC has figured prominently in this respect. The LMC has recessional velocity $270\,\mathrm{km\,s^{-1}}$ with respect to the Sun (but note that, as the LMC is a Local Group galaxy, this is not due to the expansion of the Universe); diffuse bands having $\lambda \simeq 5000$ Å are therefore shifted to the red by ~ 4.5 Å and easily identified. This is clearly seen in Fig. 9.4, which shows the 5780.4 Å band in Galaxy dust and in the LMC. Supernova

1986G erupted in the dust lane of the galaxy NGC 5128 (recessional velocity 420 km s^{-1}) and in this case also diffuse bands, shifted to the red by the Universal expansion, were detected.

The eruption of supernova 1986G also provided a rare opportunity of carrying out an investigation of the polarization properties of dust in another galaxy. The wavelength-dependence of the polarization was well-fitted by Serkowski's law [Eq. (6.53)], i.e. it was similar to that for interstellar dust grains in our own Galaxy. However, the wavelength of maximum polarization was significantly lower ($\lambda_{\text{max}} = 0.43\,\mu$m), suggesting that the ratio of total-to-selective extinction R, and hence grain size, are smaller for the grains in the dust lane of NGC 5128. Assuming that, as in our own Galaxy, the interstellar grains in NGC 5128 are aligned by its interstellar magnetic field, the position angle of the polarization of the supernova suggested that the magnetic field direction is parallel with that of the dust lane (see Fig. 9.2).

9.4.2 Infrared evidence for dust in other galaxies

Just as circumstellar dust shells around individual stars will absorb short wavelength radiation and re-emit in the infrared, we can also expect that dust particles in galaxies will emit at infrared wavelengths. This emission can arise by virtue of the fact that interstellar dust grains in the spiral arms of galaxies will (as in our own Galaxy) absorb background starlight and emit at far infrared wavelengths. The flux distribution of two nearby spiral galaxies (M31 and M33) are shown in Fig. 9.5. A similar, though more dramatic, case is provided by starburst galaxies, which are undergoing an intense phase of star formation. As a result, they contain a large number of hot, recently formed stars which heat grains in the molecular clouds in which they formed. In some cases the star-forming activity is so extreme that the galaxy is completely obscured in the optical; the existence of such objects has been confirmed by infrared surveys, particularly that carried out by *IRAS*. In the case of starburst galaxies, as in the case of normal spirals, the power source for heating the dust is not a point source but consists of the embedded, recently-formed stars which are distributed throughout the galaxy. The infrared emission from the starburst galaxies M 82 and Arp 220 is shown in Fig. 9.5. The large excess emission in the far infrared due to emission by dust is evident.

In other cases the heating of dust grains is by a powerful (usually non-thermal) source in the galactic nucleus, usually an AGN. In case of AGN the calculation of the infrared emission follows closely that of the corresponding calculation for circumstellar dust (see Section 7.3.3) except that, for the case of an AGN, the central source of radiation is likely to be non-thermal, or

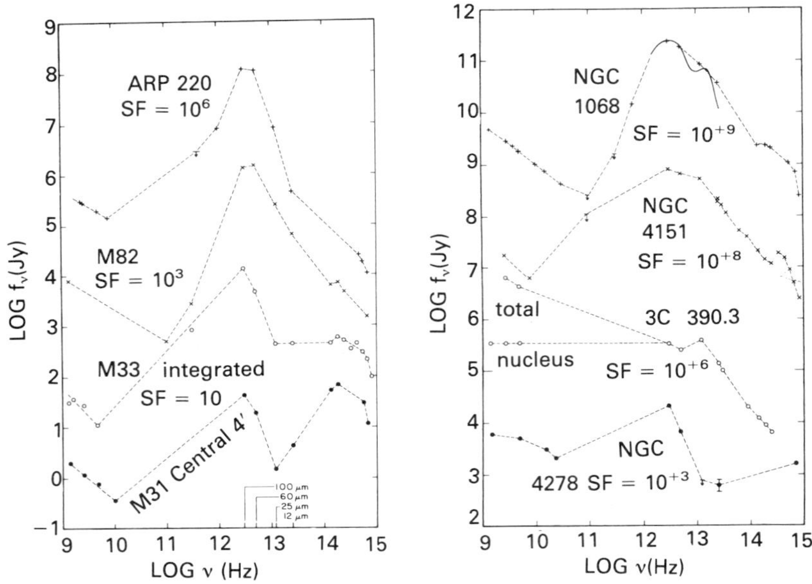

Figure 9.5: Left: The flux distributions of the starburst galaxies (Arp 220 and M 82) and two nearby spirals (M31 and M33). The plot for M33 shows the emission from the whole galaxy whereas that for M31 is for the central bulge only. The location of the *IRAS* wavebands are indicated. Right: The flux distributions of active galactic nuclei. In both figures the vertical scale is arbitrary (SF is the scale factor by which the plotted data should be divided to get actual fluxes); the dotted lines have been inserted to guide the eye. After B. T. Soifer et al. Reproduced, with permission, from the *Annual Review of Astronomy & Astrophysics*, Vol. **25**, 187. ©1987 Annual Reviews Inc.

it may be thermal emission from a hot disc surrounding the central power source. In either case the radiation that heats the 'circumnuclear' dust may, to a good approximation, be treated as a point source. Fig. 9.5 shows the infrared flux distribution of the Seyfert galaxies NGC 1068 and 4151; again the formidable emission in the far infrared, in this case due to the heating of circumnuclear dust by a source located at the nucleus of the galaxy, is evident.

Controversy sometimes arises when the nature of the galaxy can not readily be identified from optical observation, because in each case the flux distributions in the far infrared, where both starburst and AGN galaxies emit strongly, are somewhat similar (cf. Fig 9.5). However, starburst galaxies may generally be identified by the fact that their far infrared emission is extended whereas the far infrared emission from an AGN is essentially a point source.

We finally note that the far infrared emission associated with newly-formed stars might be used to identify newly-formed (i.e. the earliest) galaxies. A new galaxy is likely to have formed out of the primeval mix of hydrogen and helium. They would contain a population of young, highly luminous stars which evolve rapidly and produce the heavy elements of which dust grains are composed. Continuing star formation will heat the dust and the newly-formed (starburst) galaxy will emit strongly in the far infrared. However, galaxy formation may occur at a redshift $z \sim 10$, so that the far infrared ($\sim 100\,\mu$m) emission would be shifted into the millimetre range. This characteristic signature of galaxy formation is currently being searched for.

9.5 Intergalactic dust

The possibility that there might exist an extragalactic equivalent of interstellar dust has intrigued astrophysicists for many years. There are several reasons for this interest. For some time it has been suspected that there is a substantial amount of dark (non-luminous) matter in the Universe. One reason for this is that a detailed analysis of Hubble's Law suggested that the expansion of the Universe is decelerating. That the expansion of the Universe is decelerating is not unexpected: as each galaxy recedes it has to work against the Universal gravitational field and consequently the general expansion is slowing down. The rate at which it slows down depends on the strength of the gravitational field and hence on the density of matter in the Universe (it obviously makes no sense to speak of the mass of the Universe in this context). One can therefore relate the observed deceleration to the mean density of the Universe, which can (in principle) be determined. However the amount of visible matter in the Universe (in the form of galaxies, clusters of galaxies, quasars) is far from sufficient to account for the deceleration that was observed–hence the 'missing mass' problem. One possible solution was that the missing mass was in the form of dark, non-luminous matter; after all, the only matter that we can actually see to determine the mass density must by definition be luminous (e.g. galaxies). One candidate (amongst many others) for this dark matter was intergalactic dust.

9.5.1 Intergalactic 'footballs'

In this case, as in the interstellar case, we may be talking not only about small, sub-micron sized grains, but also about much larger particles. Indeed this question has been dubbed the 'football problem' and indeed, for all we know, intergalactic space could be filled with copies of *The Dusty Universe.* The problem with such material–matter in the form of large (dimensions $a \gg \lambda$) rather than sub-μm particles–is that there is no means at present of detecting it. For example, its extinction law, as far as visible light is concerned, is 'neutral'–i.e. independent of wavelength. This follows from the behaviour of the extinction efficiency Q_{ext} with frequency, as shown in Fig. 3.7: since we are in the region where $2\pi a/\lambda \gg 1$ the extinction is independent of wavelength. Such matter, if it existed, would serve only to render distant objects slightly less bright, equally at all wavelengths. However, the dynamical effect of such matter would be profound as it might contribute significantly to 'dark matter' in the Universe.

On the other hand, certain types of 'intergalactic football' can be ruled out without recourse to observation. We know that primeval matter consisted almost entirely of hydrogen (73 per cent) and helium (25 per cent), with a negligible amount of all heavier elements. Now in the vicinity of our own Galaxy the temperature of any object in equilibrium with the microwave background is well-determined to be $T_{\text{cos}} = 2.735\,\text{K}$, while at redshift z the corresponding temperature is $T_{\text{cos}}(1+z)$. [Note however that, in the environments of certain clusters of galaxies, there is gas at X-ray emitting temperatures ($\sim 10^6 - 10^7$ K); we are not concerned with these environments in the following discussion.]. Now if we have hydrogen at this sort of temperature, is there any possibility that some may condense out in the form of intergalactic hydrogen 'snowballs' or even 'icebergs'? We approach this problem in much the same way as we approached the formation of grains in Chapter 5 and in the spirit of our non-rigorous treatment of cosmological matters.

Suppose we have spherical hydrogen snowballs, of radius a, where a is a quantity to be determined. Such objects would absorb and scatter the light from distant galaxies and, although the resultant extinction would be neutral, it would increase with increasing distance. The extinction optical depth for large ($a \gg \lambda$) snowballs is

$$\tau_{\text{ext}} \sim n_{\text{SB}} \pi a^2 cz/H, \tag{9.4}$$

for small enough distances; here n_{SB} is the number of snowballs per unit volume in our vicinity. In fact Eq. (9.4) underestimates the optical depth because it ignores the effect of the expansion of the Universe on the number density n_{SB}. When we examine the way in which the apparent luminosity of standard galaxies decreases with distance there seems to be little evidence of extinction, suggesting that the $\tau_{\text{ext}} \lesssim 1$, at least for $z \lesssim 0.5$, in Eq. (9.4). The density of solid hydrogen is $\rho_{\text{H}} = 70\,\text{kg}\,\text{m}^{-3}$ so the mass of each individual snowball is

$$m_{\rm SB} = \frac{4\pi a^3 \rho_{\rm H}}{3}$$

and the average mass per unit volume of solid hydrogen over intergalactic space is $\rho_{\rm SB} = n_{\rm SB} m_{\rm SB}$. Thus with $\tau \lesssim 1$ in Eq. (9.4) we find

$$a > 0.5 \times \frac{3c}{4\rho_{\rm H} H} \rho_{\rm SB}$$

or, in terms of the critical density $\rho_{\rm C}$:

$$a \gtrsim 0.01 \left(\frac{\rho_{\rm SB}}{\rho_{\rm C}}\right) \ {\rm m}.$$

This tells us that, if there are intergalactic hydrogen snowballs, they would be large flakes if their density were comparable with the critical density.

To determine whether or not these snowballs can survive we need consider the likely means of destruction. Of the processes discussed in Chapter 5 the only process likely to be relevant is evaporation. We therefore need to know the vapour pressure of hydrogen, which is given by the Clausius-Clapeyron equation (5.1) and the parameters of Table 5.1. From Eq. (5.28), and using a temperature for intergalactic matter of 2.735 K, we see that the evaporation time of a hydrogen snowball is $\tau_{\rm evap} \simeq 10^7 a$ years, where the snowball radius is in metres. This is considerably less than the age of the Universe, 13×10^9 years. Thus unless the snowballs are very large–several kilometres–they would not survive long in intergalactic space and their presence can therefore be ruled out.

Now this leads us to an interesting conclusion. If there is any intergalactic dust, it must presumably be made up of pretty much the same materials as interstellar dust, i.e. silicates, carbon (and its compounds), iron (and its compounds) etc. Now we know that, from the original primeval hydrogen, only helium was formed in significant quantities in the first few minutes of cosmic time: the raw materials for grain formation (carbon, magnesium, silicon, iron etc.) were certainly not produced in the early Universe. Where then might the raw materials to form intergalactic dust have originated? They must have arisen as a result of stellar evolution, because it is only in the interiors of stars that the heavy elements required for grain formation can be formed in significant quantities. However this only pushes the question back a step: where are the stars in which the required nucleosynthesis occurred? There are two possible answers to this question. First there are the stars in galaxies which, as we have already discussed, supply their local interstellar space with dust grains. The interstellar grains could then be 'blown' out of the parent galaxy by radiation pressure, and thus fill intergalactic space with dust. In this case the rôle of galaxies with respect to intergalactic dust is analogous to the rôle of stars in relation to interstellar dust. Second, there may have existed in the early Universe a population of stars ('Population III') which formed and evolved before the galaxies themselves had formed. Such stars would have produced dust in much the same way as do stars in our own Galaxy, and ejected it into the surrounding environment. While all trace of these Population III stars has now gone, they may have left a relic of their existence in the form of intergalactic dust.

9.5.2 Search for intergalactic dust

The search for any intergalactic dust can be conducted in three ways–extinction, obscuration and infrared emission–depending on its possible location. There might be dust grains *within* clusters of galaxies, but *between* the member galaxies themselves; such dust would, in the strict sense of the word, be intergalactic and intracluster. Then there might be dust *between* clusters of galaxies–dust that is truly intergalactic.

Intracluster dust would be relatively easy to look for, if only for the fact that it is localized on the sky in the vicinity of the cluster. In this case we can investigate whether distant galaxies seen through a nearby cluster of galaxies seem fainter or redder than we would expect them to be in the absence of the cluster. Or extinction by intracluster dust may render distant (background) galaxies so faint that they are no longer visible: might there be a deficiency of faint galaxies in the direction of nearby clusters? Also in the cases of many large clusters of galaxies the presence of hot ($\sim 10^7$ K) gas is revealed by its X-ray emission. Now we know (Section 4.3.2) that dust grains in a hot gas are heated by gas impact and so some diffuse infrared emission may be expected from intracluster dust in clusters known to be sources of diffuse X-ray emission.

Similarly, in the case of truly intergalactic dust we can see whether, after allowing for extinction in our own Galaxy, there is any evidence for an additional component of extinction or reddening whose magnitude increases with increasing redshift. For example, do quasars of higher redshift appear redder than those of lower redshift? One of the problems with detecting dust grains in intergalactic space is disentangling the effects of such dust from evolutionary effects. For example, we might find evidence for the fact that more distant objects are redder than nearer ones, suggesting the cumulative effects of interstellar-like dust grains in intergalactic space. However our lack of knowledge of how galaxies and related objects change over cosmological timescales is so inadequate that such an increase in reddening could equally well be due to intrinsic changes in the colours of these objects over long timescales: recall that when we observe an object at a distance of (say) 2 Gpc we see it as it was some 6×10^9 years ago, when the properties of galaxies may have been very different from those of galaxies in our vicinity.

An alternative method would be to exploit the fact that, if intergalactic dust does indeed exist, it should scatter X-rays in much the same way as does interstellar dust. Suitable objects (like quasars) that are extremely distant and bright at X-ray wavelengths, should therefore possess detectable X-ray haloes. Such haloes have yet to be searched for.

We might also look at the distribution of galaxies on the sky: are there 'holes' in the distribution of galaxies, akin to those (originally found by Herschel) in the distribution of stars? If we look at the distribution of the nearer galaxies on the sky we do indeed find gaps and, if these gaps are due to clouds of intergalactic dust, we ought to find that they are also devoid of more distant galaxies. However when we now look at the distribution of more distant galaxies in the same areas of sky we find that they occupy the gaps between the nearer galaxies, contrary to what one would expect if the gaps were due to clouds of intergalactic dust. The situation is further complicated by the fact that, on very large scales, the 'three-dimensional' distribution of clusters of galaxies is far from uniform and consists of large 'voids', with clusters of galaxies located on long filaments. Clearly this means of searching for intergalactic dust is unlikely to be fruitful.

Problems

N.B. Neglect the expansion of the Universe in all but 9.1.

9.1. Over what redshift range will the '2200Å' extinction bump be found in the optical spectra of quasars?

9.2. Suppose that intergalactic dust has the same scattering properties as interstellar dust in our Galaxy. An X-ray emitting quasar is located at Galactic latitude $b = 90°$ so that its X-rays enter the Galactic disc at right angles. Use the arguments in Section 6.5.5 to estimate the angular dimensions of the X-ray halo due to scattering by (i) intergalactic dust, uniformly distributed along the line of sight and (ii) interstellar dust in our Galaxy.

9.3. There exists an X-ray background due to hot intergalactic gas at temperature $T_{\rm gas} = 10^7$ K and number density 10 hydrogen ions $\rm m^{-3}$. If there also exists a population of intergalactic dust grains what is their temperature due to the impact of the hot gas? Might this effect account for the cosmic microwave background?

Reading

For some superb photographs of galaxies the following can not be bettered:

Atlas of Galaxies, A. Sandage & J. Bedke, NASA publication SP-496 (1988).

The basic ideas of the expanding Universe are well described in

[A] *The Big Bang*, J. Silk, W. H. Freeman (1980).

[B/C] *Cosmology*, M. Rowan-Robinson, Oxford University Press (1976).

[C] *Principles of Cosmology and Gravitation*, M. Berry, Adam Hilger.

For a thoroughly entertaining discussion of some cosmological ideas, see

[A/B] *The First Three Minutes*, S. Weinberg, Andre Deutch (1976).

IRAS observations of galaxies, including their dust emission, is reviewed in

[D] B. T. Soifer et al., *Annual Review of Astronomy & Astrophysics*, Vol. **25**, 187 (1987).

The discussion in Section 9.5.1 is based on

[D] *Physical Cosmology*, P. J. E. Peebles, Princeton University Press (1971).

Appendix A

Equilibrium charge on a spherical grain

We look at a collision between a positive ion and a spherical grain of radius a, which carries a negative charge Ze, Z being an integer (see Fig. A1). The velocity of the ion at infinity (essentially the velocity of the ion before it begins to feel the coulomb repulsion of the grain) is v_{i}. We take an imaginary line through the grain, parallel with the ion's velocity at infinity; the distance between this line and the ion's trajectory–the *impact parameter*–is b. In this discussion we take the special case in which the ion just grazes the surface of the grain. We again take the energy conservation equation (4.5) but in addition we need to conserve angular momentum. The angular momentum at infinity is bv_{i}, while the angular momentum when the ion just touches the grain surface is $V_{\mathrm{i}}a$; thus

$$bv_{\mathrm{i}} = aV_{\mathrm{i}}. \tag{A.1}$$

Eliminating the unknown V_{i} between Eqs. (4.5) and (A.1) we get

$$b^2 = a^2\left(1 + \frac{2Ze^2}{4\pi\epsilon_0 a m_{\mathrm{i}} v_{\mathrm{i}}^2}\right). \tag{A.2}$$

In other words, the *cross-section* σ_{i} for the ion-grain collision is given by

$$\sigma_{\mathrm{i}} = \pi b^2 = \pi a^2\left(1 + \frac{2Ze^2}{4\pi\epsilon_0 a m_{\mathrm{i}} v_{\mathrm{i}}^2}\right) \tag{A.3}$$

for an ion with velocity v_{i} at infinity. Note that $\sigma_{\mathrm{i}} > \pi a^2$ because of the coulomb attraction between the ion and the grain. Similarly the cross-section for an electron-grain collision is

$$\sigma_{\mathrm{e}} = \pi a^2\left(1 - \frac{2Ze^2}{4\pi\epsilon_0 m_{\mathrm{e}} v_{\mathrm{e}}}\right), \tag{A.4}$$

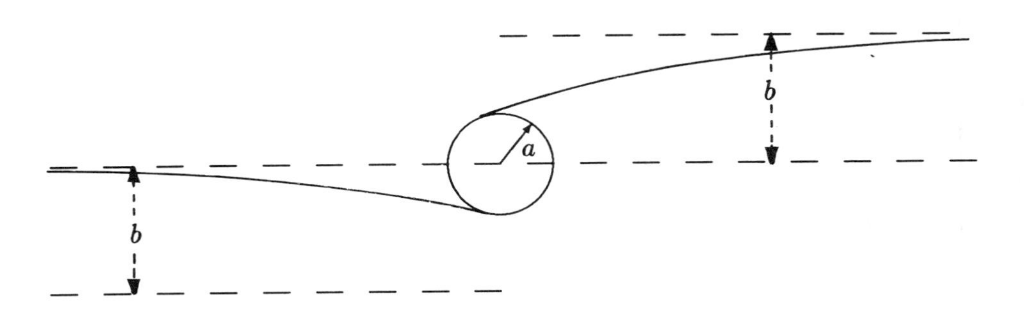

Figure A.1: Collision between an electron (left), an ion (right) and a negatively charged grain of radius a; b is the impact parameter.

where v_e is the velocity of the electron at infinity. We see that σ_e is zero for some electron velocities. For

$$v_e < v_0 = \left(\frac{2Ze^2}{4\pi\epsilon_0 a m_e} \right)^{1/2} \tag{A.5}$$

the kinetic energy of the electron is insufficient to overcome the coulomb repulsion between the electron and the grain; the electron is deflected and fails to make contact with the grain at all.

We have now determined the cross-section for electron-grain and ion-grain collisions, for electrons and ions having a specific velocity. We must now sum over all velocities and we consider the electrons first. The rate at which electrons of velocity v_e strike the grain is $\mathcal{R}_e = n_e v_e \sigma_e$, where n_e is the electron number density and σ_e is given by Eq. (A.4). However not all these electrons will stick to the grain surface; we must multiply $\mathcal{R}_e$ by the sticking probability S_e, the probability of an incident electron remaining on the grain surface. Thus the rate (in $\mathrm{C\,s^{-1}}$) at which the grain accumulates negative charge is

$$e n_e \sigma_e v_e S_e. \tag{A.6}$$

Similarly, the rate at which the grain *loses* negative charge as a result of ion collisions is

$$e n_i \sigma_i v_i S_i. \tag{A.7}$$

We now integrate over all velocities. The distribution of velocities in a gas at temperature T_{gas} is given by the Maxwell speed distribution $f(v)dv$ (see Appendix E). The total rate of loss of negative charge is given by integrating Eq. (A.7) over $f(v)dv$ as given by Eq. (E.1):

$$\dot{Q}_+ = en_{\text{i}} \int_0^\infty \sigma_{\text{i}} S_{\text{i}} v_{\text{i}} f(v_{\text{i}}) dv_{\text{i}}, \tag{A.8}$$

where the sticking probability has, for the moment, remained as part of the integrand because laboratory measurements suggest that S_{i} may depend not only on the velocity of impact but also on the angle of incidence, the local nature of the surface where the ion strikes the grain etc. However we have no detailed information on how S_{i} varies with v_{i} and it is usual to assume that it is independent of velocity. Thus

$$\dot{Q}_+ = en_{\text{i}} S_{\text{i}} \int_0^\infty \pi a^2 \left(1 + \frac{2Ze^2}{4\pi\epsilon_0 a m_{\text{i}} v_{\text{i}}^2}\right) v_{\text{i}} f(v_{\text{i}}) dv_{\text{i}}, \tag{A.9}$$

where we have substituted for σ_{i} from Eq. (A.3). Inserting the Maxwell distribution (E.1) we get

$$\dot{Q}_+ = en_{\text{i}} S_{\text{i}} \pi a^2 \left(\frac{8kT_{\text{gas}}}{\pi m_{\text{i}}}\right)^{1/2} \int_0^\infty \left(1 + \frac{2Ze^2}{4\pi\epsilon_0 a m_{\text{i}} v_{\text{i}}^2}\right) v_{\text{i}}^3 \exp\left[-\frac{m_{\text{i}} v_{\text{i}}^2}{2kT_{\text{gas}}}\right] dv_{\text{i}}. \tag{A.10}$$

We proceed by making the substitution $x = m_{\text{i}} v_{\text{i}}^2 / 2kT_{\text{gas}}$:

$$\dot{Q}_+ = en_{\text{i}} S_{\text{i}} \pi a^2 \left(\frac{8kT_{\text{gas}}}{\pi m_{\text{i}}}\right)^{1/2} \int_0^\infty \left(x + \frac{Ze^2}{4\pi\epsilon_0 a k T_{\text{gas}}}\right) e^{-x} dx. \tag{A.11}$$

Carrying out the integration we get

$$\dot{Q}_+ = \frac{8}{\pi^{1/2}} \left(\frac{kT_{\text{gas}}}{2m_{\text{i}}}\right)^{3/2} en_{\text{i}} S_{\text{i}} \pi a^2 \left(1 + \frac{Ze^2}{4\pi\epsilon_0 a k T_{\text{gas}}}\right). \tag{A.12}$$

The analysis for the rate of charging by electrons proceeds in exactly the same way, except that we must take note of the fact that some electrons will miss the grain altogether [see Eq. (A.5)] and these electrons will not contribute to the charging of the grain. In this case

$$\dot{Q}_- = en_{\text{e}} S_{\text{e}} \pi a^2 \int_{v_0}^\infty \left(1 + \frac{2Ze^2}{4\pi\epsilon_0 m_{\text{e}} v_{\text{e}}}\right) v_{\text{e}} f(v_{\text{e}}) dv_{\text{e}}, \tag{A.13}$$

where we have again assumed that S_{e} is independent of velocity. Carrying out the integration as before we find that

$$\dot{Q}_- = \left(\frac{8kT_{\text{gas}}}{\pi m_{\text{e}}}\right)^{1/2} en_{\text{e}} S_{\text{e}} \pi a^2 e^{-x_0}, \tag{A.14}$$

where

$$x_0 = \frac{Ze^2}{4\pi\epsilon_0 akT_{\mathrm{gas}}}.$$

When we have equilibrium the rates of charging (by electrons) and discharging (by ions) are the same: $\dot{Q}_- = \dot{Q}_+$. It also helps if we assume that $S_\mathrm{e} = S_\mathrm{i}$, because both sticking probabilities then drop out of the discussion. We finally find that

$$1 + x_0 = \left(\frac{m_\mathrm{i}}{m_\mathrm{e}}\right)^{1/2} \left(\frac{n_\mathrm{e}}{n_\mathrm{i}}\right) \exp[-x_0]. \tag{A.15}$$

Note that, in general, n_e will not equal n_i. Eq. (A.15) can not be solved algebraically: it is a transcendental equation that has to be solved numerically. However, given the ratio $n_\mathrm{e}/n_\mathrm{i}$, the solution of Eq. (A.15) is that x_0 is a pure number (e.g. $x_0 \simeq 2.5$ for pure hydrogen $[m_\mathrm{i} = m_\mathrm{H}]$ and $n_\mathrm{e} = n_\mathrm{i}$) and so [cf. Eq. (4.7)]

$$Z = x_0 kT_{\mathrm{gas}} \frac{4\pi\epsilon_0 a}{e^2}. \tag{A.16}$$

Reading

The derivation here follows that in:

[C/D] *Interstellar grains*, N. C. Wickramasinghe, Chapman & Hall (1967).

Appendix B

Grain size distribution from grain-grain shattering

We show here how a power-law distribution of grain sizes might arise as a result of shattering following grain-grain collisions. Since the size of individual grains is changing as a result of collisions, we must expect that the size distribution also changes with time; we therefore define $n(a,t)da$ as the number of grains per unit volume, having radii in the range $(a \rightarrow a + da)$ at time t. The probability per unit time that a collision between two grains results in shattering is defined as $\phi(a,t)dt$. We expect ϕ to depend on the frequency of collision; for a collision between two spherical grains, of radii a and A, the collision cross-section is $\sigma_{\mathrm{g-g}} = \pi(a+A)^2$. If the relative velocity of impact is $\overline{v}$ then the collision frequency for grains of radius a is obtained by integrating $\sigma_{\mathrm{g-g}}\overline{v}$ over the grain size distribution; thus

$$\phi(a,t) = \psi\overline{v}\int_{\mathrm{A}} n(A,t)\pi(a+A)^2 dA, \tag{B.1}$$

where ψ is the probability that a collision results in shattering and we have made the simplifying assumption that the probability of shattering is independent of grain size. We can further write

$$\frac{\partial n(a,t)}{\partial t} = -n(a,t)\phi(a,t). \tag{B.2}$$

Note that, in Eq. (B.2), we have included only the removal of larger grains from the upper end of the size distribution by shattering; the *addition* of the shattered fragments to the lower end of the size distribution has been omitted. If, in the shattering of a grain of radius A, a fraction of the fragments having radii in the range $A \rightarrow A + dA$ is $\chi(a,A)$, we can complete Eq. (B.2) by the addition of the term

$$\int_a^{a_{\max}} n(A,t)\phi(A,t)\chi(a,A)dA.$$

The differential equation that governs the changing size distribution is therefore

$$\frac{\partial n(a,t)}{\partial t} = -n(a,t)\phi(a,t) + \int_a^{a_{\max}} n(A,t)\phi(A,t)\chi(a,A)dA. \tag{B.3}$$

Eq. (B.3) may be solved for $n(a,t)$ for the simple case where we consider only the removal of grains by shattering and neglect the contribution of the shattered fragments to $n(a,t)$; such an approximation is valid only at the upper end of the size distribution but even so, we will find that the resulting size distribution is one that agrees reasonably well with observational data. In this case the positive term on the right hand side of Eq. (B.3) is neglected and the equation, together with Eq. (B.1), may be written as

$$\frac{\partial n(a,t)}{\partial t} = -n(a,t)\psi\overline{v}\int_A n(A,t)\pi(a+A)^2 dA. \tag{B.4}$$

If we suppose that a steady state has been attained we can solve Eq. (B.4) by the usual technique of separation of variables, i.e. we suppose that

$$n(a,t) = f(a)g(t), \tag{B.5}$$

where f is a function of a only, and g a function only of t. Thus

$$\frac{\partial n(a,t)}{\partial t} = f(a)\frac{dg}{dt},$$

since f is independent of t. Substituting in Eq. (B.4) we get

$$\frac{dg}{dt} = -[g(t)]^2\psi\overline{v}\int_A f(A)\pi(a+A)^2 dA; \tag{B.6}$$

note that we have taken $g(t)$ outside the integration because as far as the integration over A is concerned it may be treated as a constant. Rearranging Eq. (B.6) we find

$$\frac{1}{[g(t)]^2}\frac{dg}{dt} = -\psi\overline{v}\int_A f(A)\pi(a+A)^2 dA. \tag{B.7}$$

Note that the left hand side of Eq. (B.7) depends on t only, while the right hand side depends only on a; this can only be the case if both sides are equal to a constant, the separation constant, which is independent of both a and t. We denote this constant, by $-\kappa^2$:

$$-\psi\overline{v}\int_A f(A)\pi(a+A)^2 dA \quad = \quad -\kappa^2 \tag{B.8}$$

$$\frac{1}{[g(t)]^2}\frac{dg}{dt} \quad = \quad -\kappa^2. \tag{B.9}$$

The differential equation in g is easily solved to give

$$g(t) = \frac{g_0}{1+g_0\kappa^2 t}, \tag{B.10}$$

where g_0 is the value of g at $t=0$.

We have not yet specified the limits of integration in Eq. (B.8). The integration is a measure of the collision frequency for grains of radius a. The appropriate upper limit is therefore the largest grain size in the distribution, $a_{\max}$. As we have already noted, we are confining our attention to the upper end of the size distribution and for a grain of radius a we include collisions only with grains whose radii are a specified fraction of a, say ηa.

The equation in f in Eq. (B.8) therefore gives

$$\overline{v}\pi\psi\int_{\eta a}^{a_{\max}} f(A)[a+A]^2 dA = \kappa^2 \tag{B.11}$$

from which we have to extract $f(a)$. We proceed by making the change of variable $x = A/\eta$ in the integral, so that Eq. (B.11) becomes

$$\psi\overline{v}\pi a^2\eta\int_a^{a_{\max}/\eta} f(\eta x)\left[1+\frac{\eta x}{a}\right]^2 dx = \kappa^2. \tag{B.12}$$

Now differentiate Eq. (B.12) with respect to a, and noting the dependence of the lower limit of integration on a:

$$f(\eta a)(1+\eta)^2 + \frac{2\eta}{a^2}\int_a^{a_{\max}/\eta} x f(\eta x)\left[1+\frac{\eta x}{a}\right] dx = \frac{2\kappa^2}{\overline{v}\eta\psi\pi a^3}.$$

With a little rearrangement we get

$$f(a\eta) = \frac{2\kappa^2}{\overline{v}\eta(1+\eta)^2\psi\pi a^3} - \frac{2\eta}{(1+\eta)^2 a^2}\int_a^{a_{\max}/\eta} x f(x\eta)\left[1+\frac{\eta x}{a}\right] dx \tag{B.13}$$

Now Eq. (B.13) is an integral equation which can be solved using standard techniques. However all we need note is that, as η is allowed to become smaller (without violating our assumption that we are at the upper end of the size distribution), the second term on the right hand side ($\propto \eta$) becomes negligible while the first term ($\propto \eta^{-1}$) becomes dominant. Therefore $f(a)$ gets closer to

$$f(a) \simeq \frac{2\kappa^2}{\overline{v}\eta\psi\pi a^3}. \tag{B.14}$$

Our final result is thus

$$n(a,t) \simeq \frac{g_0}{1 + g_0\kappa^2 t}\frac{2\kappa^2}{\overline{v}\eta\psi\pi}a^{-3}. \tag{B.15}$$

We see from Eq. (B.15) that the number of particles in a given size range ($a \rightarrow a + da$) decreases with time t; this is to be expected because, in confining our attention to the upper end of the size distribution, we have taken into account the *removal* of grains by shattering but we have not *added* the shattered fragments lower down the size distribution. The important feature of Eq. (B.15), however, is that it predicts $n(a)da \propto a^{-3}da$, as required. Furthermore, if we were to add the shattered fragments the effect would be to increase the number of small grains, and hence steepen the size distribution, as observed.

Reading

The treatment in this Appendix is based on the following paper;

[D] B. Hellyer, *Monthly Notices of the Royal Astronomical Society*, Vol. **148**, 383 (1970).

Appendix C

Partition function for a PAH molecule

In this Appendix we show how the partition function for a PAH molecule can be inverted to give the density of states. The partition function Z for a system at temperature T is defined in the usual way by

$$Z = \sum_{\mathrm{i}} g_{\mathrm{i}} \exp[-E_{\mathrm{i}}/kT], \tag{C.1}$$

where g_{i} is the statistical weight of level i. Where the energy levels are closely spaced the sum in Eq. (C.1) can be replaced by an integral:

$$Z = \int_0^E \rho(E) \exp[-E/kT] dE, \tag{C.2}$$

where the statistical weight in Eq. (C.1) has been replaced by the density of states function $\rho(E)$.

We suppose now that the partition function in the form (C.2) may be approximated by

$$Z = \int_0^\infty \rho(E) \exp[-\beta E] dE, \tag{C.3}$$

where $\beta = 1/kT$. The expression for Z in Eq. (C.3) may be recognized as the Laplace transform of $\rho(E)$, i.e.

$$Z = \mathcal{L}[\rho(E)] \tag{C.4}$$

and the density of states function may simply be obtained by taking the inverse transform.

We note from Eq. (8.1) that the partition function for a PAH molecule that for a collection of n independent harmonic oscillators:

$$Z = \frac{(kT)^n}{\prod_{i=1}^n h\nu_i}. \tag{C.5}$$

herefore

$$\mathcal{L}[\rho(E)] = \frac{(kT)^n}{\prod_{i=1}^n h\nu_i}$$

and

$$\rho(E) = \mathcal{L}^{-1}\left(\frac{(kT)^n}{\prod_{i=1}^n h\nu_i}\right) = \frac{E^{n-1}}{(n-1)!\prod_{i=1}^n h\nu_i}.$$

This is the result quoted in Eq. (8.2).

Reading

The above derivation is from:

[D] W. Forst, *Theory of unimolecular reactions*, Academic Press (1973).

Appendix D

Astronomical co-ordinate systems

The location of an object on the sky requires two numbers, analogous to geographic longitude and latitude. The co-ordinates chosen depend on the problem at hand: for Solar System studies it is often convenient to use a co-ordinate system based on the plane of the Earth's orbit, whereas for studies of interstellar matter it is often convenient to use a system based on the plane of the Galaxy. As these co-ordinate systems have been referred to occasionally we describe them briefly here.

D.1 The celestial equator

For the purpose of establishing systems of co-ordinates the Earth is regarded as being at the centre of a large imaginary sphere (the celestial sphere), on the inner surface of which are the 'fixed' stars, planets etc. The projection of the Earth's equator onto the celestial sphere defines the *celestial equator*, which serves as the basis for the equatorial system of co-ordinates. Although the details of equatorial co-ordinates need not concern us here the celestial equator is used in the definition of other co-ordinate systems, as discussed below. It follows that the north and south celestial poles lie directly over the geographic north and south poles respectively.

D.2 Ecliptic co-ordinates (λ, β)

As a consequence of the Earth's orbital motion, the Sun appears to move with respect to the background stars and returns to the same point on the sky after a period of one year[1]. The path of the Sun around the sky defines the *ecliptic*; obviously, the plane of the ecliptic coincides with the plane of the Earth's orbit. The *ecliptic latitude* β is measured in degrees north (+) or south (−) of the ecliptic.

Since the plane of the Earth's equator is inclined (by about 23°.5) to the plane of its orbit the celestial equator is inclined (by the same amount) to the ecliptic; the two points at which the ecliptic cuts the equator are known as the equinoxes. The point at which the Sun crosses the celestial equator from south to north is called the vernal equinox and this marks the point on both the equator and the ecliptic from which longitude is measured. *Ecliptic longitude* λ is measured around the ecliptic in degrees, with $\lambda = 0$ at the vernal equinox.

D.3 Galactic co-ordinates (l, b)

In the Galactic system of co-ordinates the plane of the Galaxy (see Section 9.2) defines the *Galactic equator* with respect to which Galactic latitude b is measured in degrees north (+) or south (−) of the Galactic equator. In this case the zero point for the longitude is taken to be the direction of the centre of the Galaxy, which lies in the direction of the constellation Sagittarius. Galactic longitude l is measured (in degrees) around the Galactic equator.

[1]Strictly speaking, this is only one of several definitions of the 'year'; this defines the sidereal year.

Appendix E

Useful data

E.1 Constants

E.1.1 Physical constants

a	Radiation constant	7.57×10^{-16} W m^{-3} K^{-4}
c	Velocity of light *in vacuo*	2.99792×10^{8} m s^{-1}
e	Electron charge	1.60×10^{-19} C
G	Gravitation constant	6.67×10^{-11} m^3 kg^{-1} s^{-2}
h	Planck constant	6.625×10^{-34} J-s
k	Boltzmann constant	1.36×10^{-23} J K^{-1}
$m_{\rm e}$	Electron mass	9.11×10^{-31} kg
$m_{\rm p}$	Proton mass	1.67×10^{-27} kg
$r_{\rm e}$	Classical electron radius $= e^2/4\pi\epsilon_0 mc^2$	2.82×10^{-15} m
$\Re$	Gas constant	8.31 J mole^{-1} K^{-1}
σ	Stefan-Boltzmann constant	5.65×10^{-8} W m^{-2} K^{-4}
$\sigma_{\rm T}$	Thomson cross-section	6.65×10^{-29} m^2
eV	Electronvolt	1.60×10^{-19} J

E.1.2 Astronomical constants

AU	Astronomical unit	1.50×10^{11} m
pc	Parsec	3.09×10^{16} m
$M_\odot$	Solar mass	1.99×10^{30} kg
$L_\odot$	Solar luminosity	3.83×10^{26} W
$[M_{\rm bol}]_\odot$	Absolute bolometric magnitude of the Sun	+4.75

E.2 Useful Formulae

E.2.1 Blackbody formulae for temperature T

Planck function

$$B_\nu = \frac{2h\nu^3}{c^2}\frac{1}{\exp[h\nu/kT]-1}$$

$$B_\lambda = \frac{2hc^2}{\lambda^5}\frac{1}{\exp[hc/\lambda kT]-1}$$

Power emitted per unit area from the surface of a blackbody

$$\mathcal{P} = \sigma T^4$$

Energy density of radiation in a blackbody cavity

$$u_\lambda = \frac{8\pi hc^3}{\lambda^5}\frac{1}{\exp[hc/\lambda kT]-1}$$

E.2.2 Maxwell speed distribution

In a gas at temperature $T_{\rm gas}$, containing $n_{\rm X}$ atoms or molecules of X per unit volume, the *fraction* of atoms or molecules having speed v in the range $v \to v + dv$ is

$$f(v)dv = \left(\frac{2}{\pi}\right)^{1/2}\left(\frac{m_{\rm X}}{kT_{\rm gas}}\right)^{3/2} v^2 \exp\left[-\frac{m_{\rm X}v^2}{2kT_{\rm gas}}\right] dv. \tag{E.1}$$

E.2.3 Mathematical formulae

$$\int_0^\infty x^n e^{-x} dx = \Gamma(n+1) = n!$$

$$\int_0^\infty x^{n-1/2} e^{-x} = \frac{\sqrt{\pi}}{2^n}(2n-1)!! \quad \text{(!! denotes product over odd } n \text{ only)}$$

$$\int_0^\infty \frac{x^{n-1}dx}{e^x - 1} = (n-1)!\zeta(n)$$

$$\zeta(n) = \sum_{k=1}^{\infty}\frac{1}{k^n}$$

$$\zeta(4) = \pi^4/90 \simeq 1.0823\ldots$$

$$\zeta(5) \simeq 1.0369\ldots$$

$$\zeta(6) \simeq 1.0173\ldots$$

Reading

The text by Stephenson is superb and ideally complements the present volume:

[B/C] *Mathematical Methods for Science Students*, G. Stephenson, Longmans (1988).

The following has been the indispensable companion of every astronomer and astrophysicist for nearly half a century:

[C/D] *Astrophysical Quantities*, C. W. Allen, Athlone Press, London (1973).

See also

[C/D] *Handbook of Space Astronomy & Astrophysics*, M. V. Zombeck, Cambridge University Press (1990).

Index

Absolute magnitude, 13–14
Absorption of radiation, 31, 46–47, 60, 65–69, 80
Accretion disc, 167, 208
Active galactic nuclei (AGN), 159, 199, 207–208
Amorphous grains, 94, 161
  carbon, 84, 191
  silicate, 163–166
Anomalous diffraction, 130–131
Asteroid belt, 80, 98, 99
Asymptotic grain size, 92–93, 172
Atmospheric transmission, 10, 22, 26, 27
Atomic clusters, 3, 179
Atomic hydrogen, 13, 20, 27

Bipolar outflows, 127–129, 167
Broadband filters, 9–11, 104
  infrared, 10–11, 22, 24
  two colour diagram, 155–157
  UBVRI, 9–11
  wavelength response, 10–11
Buckminsterfullerene, 194–195

Carbon grains, 53, 54, 62, 84–85, 90, 138, 154, 162, 167, 168, 171, 176, 182, 211
  '2175' feature, 53, 108–110, 117
  Chemisputtering, 97–98
  Planck mean, 68
Carbon-rich stars, 16, 88, 89, 167, 172
Chemical reactions,
  grain surfaces, 149
  heating effects of, 74–76
Chemisputtering, 97–98
Circumstellar dust, 14, 23, 26, 105, 106, 153–178
  composition, 162–168
  dust mass, 154
  flux distribution, 159–161
  formation, 168–172
  luminosity, 158–159
  reddening, 158
  scattering properties, 158
  temperature, 153–154
Colour excess, 13, 105–108, 146–147
Colour indices, 12–13, 105
Complex refractive index, 50, 53, 54
Condensation distance, 128, 171–172, 175
Condensation temperature, 89, 90, 163, 171
Co-ordinate systems, 225–226
Core-mantle grains, 117, 148–151
Cosmic microwave background, 137, 201, 202–203, 210
Cross-section
  absorption, 47, 60, 67, 80, 114, 189
  extinction, 47, 111, 130
  ionization, 189
  scattering, 47
    differential, 129–131, 133–134
  collision
    electron-grain, 61, 63, 215
    ion-grain, 63, 72, 215
Crystalline grains, 94, 161
  carbon, 84
  silicate, 164, 166
Curie's law, 144

Dark matter, 209, 210
Definitions,
  albedo (ϖ), 47

Astronomical unit (AU), 4, 227
column density, 104
degree of polarization (p), 140
emission coefficient (ϵ_ν), 31
equivalent width, 35
extinction coefficient (κ_ν), 31
flux (S), 5
density (S_ν), 5
intensity (I_ν), 30
Jansky, 5
light year, 4
Lyman continuum photon, 27
optical depth (τ_ν), 31
parsec (pc), 4, 227
Planck mean, 67
redshift (z), 201
scattering phase function $S(\theta)$, 47–48
solid angle ($d\Omega$), 7
source function ($\mathcal{S}_\nu$), 31
steradian, 6
Depletion, 88, 91–94, 113–115, 117, 148, 172
Dielectric function, 44–46, 50
Diffuse interstellar bands, 26, 35, 115–116, 187–188, 206–207
Distances
distance modulus, 14
of stars, 1
units, 4
Dust grains
charge, 60–65, 215–218
photoelectric effect, 60–61
plasma, 62–65
thermionic effect, 61–62
composition, 26, 53–54, 162–168, 211
emissivity, 22, 23, 67
evaporation, 23, 95, 211
formation, 147–148
non-spherical, 55–57, 135, 142–146
surface reactions, 74–76, 149
temperature, 59, 65–78
Echo,
infrared, 176
light, 121–127
Equation,
Clausius-Clapeyron, 85
Richardson-Dushman, 62
Sackur-Tetrode, 85
Transfer, 29–37
applications, 104, 118, 133, 138
solution, 32–34
Equipartition theorem, 63, 143
Equivalent width, 34–36, 113–114
Evaporation, 95
temperature, 150
timescale, 95, 211
EXAFS, 135–136
Extended red emission (ERE), 186–187
Extinction of radiation, 46–47
elongated grains, 55–57, 142–143
interstellar, 103–110
Extragalactic dust, 197–214

Galaxies, 197–200
clusters, 200
dust in, 203–209
elliptical, 197–198
Hubble sequence, 197–198
infrared emission, 207–209
irregular, 198
Magellanic Clouds, 124–125, 200
spiral, 198, 207–208
starburst, 199, 207–209
Galaxy (Milky Way), 2, 19–20, 116, 143
Gibbs free energy, 84, 87
Grain composition
(see also: Carbon, Ice, Silicate, Silicon Carbide)
cohenite (Fe_3C), 90
corundum (Al_2O_3), 90, 162
hydrogenated amorphous carbon, 167, 187, 191
magnesium oxide (MgO), 162
magnesium sulphide (MgS), 163
magnetite (Fe_3O_4), 90, 162

polyformaldehyde ($[H_2CO]_n$), 163
silica (SiO_2), 162
silicon nitride (Si_3N_4), 163
Grain destruction, 72, 95–100
electrostatic tensile stress, 100
evaporation, 95
field ion emission, 99
shattering, 98–99, 219–222
sputtering, 96–98
Grain formation, 83–94, 168–172
Grain growth, 72, 89–91, 170–172
Grain nucleation, 86–89
heterogeneous, 88–89
homogeneous, 86–88
Grain surface,
potential well, 70, 94
reactions, 74–76
Grain-grain collisions, 98–99, 117, 219–222

HII regions, 21, 76, 113, 180, 186, 192, 193, 205
Hertzsprung-Russell (H-R) diagram, 16–18, 167
Hubble's law, 200–202, 209
Hubble's relation, 118–120
Hydrogen (solid), 85, 210–211

Ice grains, 85, 90, 97, 113, 115, 149, 163
3.1 μm feature, 113, 115, 163
Individual objects
Arp 220, 207, 208
B338 (Bok globule), 24, 25
BF Ori, 140
comet Halley, 166
comet Kohoutek, 166
DQ Her, 173
Egg nebula, 127, 129
GK Per, 125
GX 9 + 9, 134–135
GX 17 + 2, 134–135
HD 31726, 108–109
HD 37367, 108–109
HD 44179, 186
Large Magellanic Cloud (LMC), 73, 124–125, 200, 203, 205, 206
LW Ser, 173
M31, 200, 207, 208
M33, 207, 208
M82, 207, 208
M100, 175
N49, 73–74
NGC 1068, 208
NGC 2327, 186
NGC 4151, 208
NGC 5128, 204, 207
NGC 7027, 187, 188
NQ Vul, 173, 174
Orion nebula, 6–7, 8, 9, 193
PC1247+3406, 201
Pleiades, 24, 25
Red rectangle, 186, 187
RS Pup, 126–127
RW Cyg, 166
RY Sgr, 176–177
Small Magellanic Cloud (SMC), 200, 203, 205
supernova 1979C, 175
supernova 1986G, 204, 207
supernova 1987A, 124-125, 175, 206
V1668 Cyg, 173
WW Vul, 168, 169
X Her, 164, 166
θ Ori, 164
Infrared cirrus, 180–181
Infrared excess, 105, 155–158, 168
Infrared spectral features, 23, 163, 187, 188, 189, 192
Intergalactic dust, 209–213
extinction law, 204–206, 210
Interplanetary dust, 80, 89, 100, 164, 166
meteorites, 174, 176, 195
Interstellar dust grains,
alignment, 143–147
composition, 111–117
dimensions, 39, 46, 111, 112, 129, 143, 147, 148, 207

extinction law, 103–110
formation, 147–148
infrared emission, 136–139
non-spherical, 55-57, 135, 142–147
PAH, 180–181, 187–188, 192
polarization due to, 139–147
reddening, 147
temperature, 136–139, 144
Interstellar extinction
absorption features, 192, 205
'2175' feature, 108–110, 115, 117, 205
'2500' feature, 192
silicate feature (9.7 μm), 115, 116
other galaxies, 204–206
wavelength-dependence, 2, 24, 106–110, 111, 147
Interstellar gas, 19–21, 34–35
absorption by, 132
composition, 21, 113–115, 205
HI, 21
HII, 21, 205
ionized gas, 21
temperature, 20, 28, 69, 73, 76, 143–144, 148
Interstellar medium
Faraday rotation, 21
free electrons, 20, 21, 111
gas-to-dust ratio, 113
magnetic field, 20, 21, 143–146
molecules, 20
Oort limit, 20, 111–112
synchrotron radiation, 21, 143
Interstellar polarization, 139–147
Serkowski's law, 141, 207
wavelength-dependence, 24, 56, 140–143, 207
Interstellar reddening, 13, 105, 106–108, 139, 146–147, 158

Kramers-Kronig relations, 45, 145

Magnetic field, 20, 143–146
Zeeman effect, 21
Magnitude scale, 8–14
absolute, 13–14
bolometric correction, 14
colour index, 12–13
infrared, 10–11
UBVRI, 9–11
Maxwell speed distribution, 71, 72, 217, 228
Meteorites (see Interplanetary dust)
Mie theory, 49–53, 189
Molecular clouds, 21, 147, 149
Molecules
(see also: PAH)
C_2, 16, 89
chrysene, 182, 188
CN, 16, 174
CO, 21, 88, 127, 162, 167
coronene, 182, 188
H_2, 21, 75–76
hydrocarbons, 88, 89, 97, 163
Interstellar gas, 20, 114–115, 179
Rayleigh scattering by, 112
MRN model, 117, 181

Novae, 19, 125, 173–175, 192, 195
Nucleation, 168, 170
Heterogeneous, 86–88
Homogeneous, 88–89
Rate (carbon), 88
Size of nucleation site, 87, 192

Oort limit, 20, 111–112

Paramagnetic relaxation, 143–146, 168
Phase diagram, 83–84
Phonon
modes
acoustic, 39–41
dispersion relation, 41, 42, 43
forbidden band, 44, 167
optical, 41–44
velocity, 44, 65
Photoelectric effect, 60–61, 64
Planck mean absorption efficiency, 67–69, 138, 159
carbon, 68
Polarization
elongated grains, 55–57

interstellar grains, 139–147
scattering, 53, 55
wavelength-dependence, 140–143, 147
Polycyclic Aromatic Hydrocarbons (PAH), 163, 179–196
charge, 189–190
clusters, 190–191
dimensions, 188–190
fluorescence, 183–184
infrared emission features, 187, 189
internal conversion, 183–184
laboratory studies, 194
partition function, 184–185, 223–224
temperature, 184–186
Population I stars, 18, 20, 198
Population II stars, 18, 20, 198
Population III stars, 18, 212
Poynting-Robertson effect, 79–81
Pre-main sequence stars, 17, 18, 154, 167–168
Herbig stars, 168, 169
T Tauri stars, 168
Properties of solids
electrical properties,
dielectric function, 44–46, 50
electrical conductivity, 46
polarizability, 44–45
magnetic properties, 144–146

Radiation pressure, 48–49, 80–81, 211
Ratio of total-to-selective extinction, 106–107, 147, 207
Rayleigh scattering, 53, 112
Red giant stars, 16, 164
Reddening, 13, 105, 106–108, 139, 146–147, 158
Redshift, 201
Reflection nebulae, 24, 117–120, 127–129, 180, 186, 191
compact, 127
Egg nebula, 127, 129
Pleiades, 24, 25

Satellite observatories,
ANS, 26
Ariel, 27
COBE, 203
Copernicus, 26
Einstein, 27, 129, 134, 135
EXOSAT, 27, 129, 134
IRAS, 23-2-4, 73–74, 158, 164, 165, 180, 207, 208
IUE, 26, 107
Uhuru, 27, 129
Scattering, 24–25, 31, 46–47, 117–135
back, 48, 55
background starlight, 26, 120–121
circumstellar dust, 158
forward, 47, 52
Heyney-Greenstein function, 52, 126
isotropic, 48, 52
phase function, 47–48, 121, 126, 129
polarization, 53, 55, 129
Rayleigh, 53, 112
Thomson, 111, 203
variable star (light echo), 121–127
X-rays, 27, 129–134, 213
Serkowski's law, 141, 207
Silicate grains, 53, 54, 62, 85, 89, 90, 115, 117, 130, 138, 148, 162, 163, 168, 211
9.7 μm feature, 53, 54, 115, 163, 164, 187
18 μm feature, 53, 54, 115, 163, 164
amorphous, 164, 166
crystalline, 164
Silicon carbide grains, 88, 90, 115, 162, 163, 167, 168
11.5 μm feature, 163, 165, 167
Size distribution, 98–99, 116–117, 134, 219–222
Small grains, 53, 69, 86, 112
Specific heat
C_V, 77
Debye theory, 77, 185
Dulong-Petit law, 77

Einstein theory, 185
PAH molecule, 185
Sputtering, 69, 96–98
Stars,
bolometric luminosity, 14
carbon-rich, 16, 167, 176
clusters, 1
distances, 1
dust-forming, 148, 170–177
early, 17
evolution, 16–18
evolved, 17, 18
giants, 18
hydrogen-deficient, 176
late, 17
luminosity class, 18
oxygen-rich, 164
populations, 18,
pre-main sequence, 17, 18, 167
red giants, 16
spectral classification, 15
variable (see Variable stars)
Stellar wind, 14, 78, 153, 160, 170–172
Sticking probability, 62, 71, 75, 91, 216
Stochastic heating, 76–78, 181, 182
Stromgren spheres, 21
(see also: HII regions),
Sun, 4–5, 8, 15, 17, 18, 79, 113, 153, 168, 202
Supernovae, 19, 28, 69, 73, 83, 124–125, 175–176, 204, 207

Temperature
colour, 13, 15, 155, 180
dust grains
chemical reactions, 74–76
condensation, 89, 171
gas impact, 69–74, 212
non-equilibrium, 76–78
radiation, 65–69, 171
effective, 15, 157
gas, 20, 28, 69, 73, 76, 143–144, 148, 210, 212
Thermionic effect, 61–62, 65
Thermonuclear processes, 162, 174, 175, 202, 211
Threshold
mantle formation, 149–151
sputtering (physical), 96
Timescale
'age' of Universe, 201
equilibrium charge, 64
grain alignment, 145
grain cooling, 77
grain evaporation, 95, 211
grain formation, 148
grain growth, 94
grain spin-up, 144
ion on grain surface, 70
surface migration, 94

Units, 4–5, 8, 227
Universe
'age', 201
critical density, 201, 211
mean density, 201, 209
scale, 4, 201

Variable stars
cepheid variables, 19, 126–127
designation, 19
novae, 19, 125, 173–175, 176
pre-main sequence, 168
R Coronae Borealis, 176–177
RV Tauri, 19, 156, 157
supernovae, 19, 28, 69, 73, 83, 124–125, 175–176, 204, 207
Viscous drag, 78–79

Wien displacement law, 15, 67
work function, 60, 61, 62, 179

X-ray scattering, 27, 129–135, 213

Yield
photoelectric, 60
photoionization, 190
sputtering (physical), 96